THE BEST TEST PREPARATION FOR THE
ADVANCED PLACEMENT
EXAMINATION IN

MATHEMATICS
CALCULUS AB

Donald E. Brook
Mathematics Instructor
Mount San Antonio College, Walnut, California

Donna M. Smith
Mathematics Instructor
American River College, Sacramento, California

Tefera Worku
Mathematics Instructor
SUNY–Albany, Albany, New York

 Research and Education Association
61 Ethel Road West
Piscataway, New Jersey 08854

The Best Test Preparation for the
ADVANCED PLACEMENT EXAMINATION
IN MATHEMATICS: CALCULUS AB

Printed in the United States of America

Library of Congress Catalog Card Number 95-69082

International Standard Book Number 0-87891-646-6

Research & Education Association
61 Ethel Road West
Piscataway, New Jersey 08854

REA supports the effort to conserve and
protect environmental resources by
printing on recycled papers.

CONTENTS

PREFACE v
ABOUT THE TEST v
ABOUT THE REVIEW SECTION vi
SCORING THE TEST vi

AP CALCULUS AB COURSE REVIEW

CHAPTER 1 - ELEMENTARY FUNCTIONS
Properties of Functions...A-1
Properties of Particular Functions A-8
Limits...A-11

CHAPTER 2 - DIFFERENTIAL CALCULUS
The Derivative A-14
Application of the Derivative A-26

CHAPTER 3 - INTEGRAL CALCULUS
Antiderivatives...A-36
Application of Antiderivatives $y=y_0e^{Kt}$:
    The Law of Exponential Change A-37
Techniques of Integration..A-38
The Definite Integral A-43
Applications of the Integral A-47

SIX PRACTICE EXAMS

AP CALCULUS AB EXAM I
Test I 1
Answer Key 22
Detailed Explanations of Answers 23

AP CALCULUS AB EXAM II
Test II 47
Answer Key 68
Detailed Explanations of Answers 69

AP CALCULUS AB EXAM III
Test III 101
Answer Key 124
Detailed Explanations of Answers 125

AP CALCULUS AB EXAM IV
Test IV 153
Answer Key 175
Detailed Explanations of Answers 176

AP CALCULUS AB EXAM V
Test V 209
Answer Key 233
Detailed Explanations of Answers 234

AP CALCULUS AB EXAM VI
Test VI 271
Answer Key 293
Detailed Explanations of Answers 294

ANSWER SHEETS 325

PREFACE

This book provides an accurate and complete representation of the Advanced Placement Examination in Calculus AB. The six practice tests and the review section are based on the most recently administered AP Calculus AB Exams. Each test is three hours in length, and includes types of questions that can be expected to appear on the actual exam. Following each test is an answer key complete with detailed explanations. The explanations discuss the correct responses and are designed to clarify the material for the student.

By studying the review section, completing all six tests, and studying the answer explanations, students can discover their strengths and weaknesses and thereby prepare themselves for the actual AP Calculus AB Examination.

Teachers of Advanced Placement Calculus AB courses will also find this book of great use. The book can be used as supplemental text in planning courses and practicing for the exam. It will provide teachers with specific information concerning the level, scope, and type of material found on the actual exam.

ABOUT THE TEST

The Advanced Placement Calculus Examination is offered each May at participating schools and multi-school centers throughout the world.

The Advanced Placement Program is designed to allow high school students to pursue college-level studies while attending high school. The participating colleges, in turn, grant credit and/or advanced placement to students who do well on the examination.

The Advanced Placement Calculus AB course is designed to represent college-level mathematics, and is intended for students who have a strong background in college-preparatory mathematics, including algebra, axiomatic geometry, trigonometry, and analytic geometry.

The exam is divided into two sections:

1) **Multiple Choice:** composed of approximately 40 multiple-choice questions in two sections, designed to measure the student's abilities in a wide range of mathematical topics. These questions vary in difficulty and complexity. Part A consists of 25 questions for which a calculator cannot be used. Part B consists of 15 questions, one-third of which may require the use of a graphing calculator. Ninety minutes is allowed for this section of the exam.

2) **Free-response Section**: composed of six free-response questions that test how well and how accurately the student is able to recall and utilize knowledge of calculus. Between 0 and 9 points are

awarded for each question based on the work shown and whether or not the solution given is correct. Partial credit is given for answers that are correct in format, yet incorrect in the solution. Therefore, it is strongly recommended that all work is written down. One hour and thirty minutes is allowed for this section of the exam. Each of these two sections counts for 50% of the student's total exam grade. Because the exam contains such a vast quantity and variety of material, students are not expected to be able to answer all the questions correctly.

The list of approved calculators for the test includes:

Casio
FX-6000, 6200, 6300, 6500, 7000, 7500, 7700,
8000, 8500, 8700, 9700

Sharp
EL-5200, 9200, 9300

Texas Instruments
TI-81, 82, 85

Radio Shack
EC-4033, 4034

The sample tests in this book provide calculator questions with explanations which include the steps necessary with the calculator. These steps are illustrated after the word "CALCULATOR" listing the key strokes that should be used.

ABOUT THE REVIEW SECTION

This book contains review material that students will find useful as a study aid while preparing for the AP Calculus AB Examination. By reviewing each item, students will be able to become better acquainted with information that will most likely appear on the actual test. Included in this section are the following topics:

Elementary Functions — This chapter describes the Properties of Functions, the Properties of Particular Functions, and Limits

Differential Calculus — This chapter deals with Derivatives and Application of the Derivative

Integral Calculus — This chapter explains Anti-Derivatives, Applications of Anti-Derivatives, The Law of Exponential Change, Techniques of Integration, The Definite Integral, and Applications of the Integral

SCORING THE TEST

SCORING THE MULTIPLE-CHOICE SECTION

For the multiple choice section, use this formula to calculate your raw score:

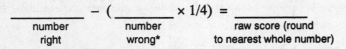

$$\underset{\substack{\text{number}\\\text{right}}}{\underline{\hspace{2cm}}} - (\underset{\substack{\text{number}\\\text{wrong*}}}{\underline{\hspace{2cm}}} \times 1/4) = \underset{\substack{\text{raw score (round}\\\text{to nearest whole number)}}}{\underline{\hspace{2cm}}}$$

* DO NOT include unanswered questions

SCORING THE FREE-RESPONSE SECTION

For the free-response section, use this formula to calculate your raw score:

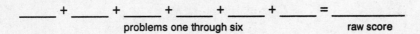

$$\underline{\hspace{1cm}} + \underline{\hspace{1cm}} + \underline{\hspace{1cm}} + \underline{\hspace{1cm}} + \underline{\hspace{1cm}} + \underline{\hspace{1cm}} = \underline{\hspace{2cm}}$$
$$\underset{\text{problems one through six}}{} \qquad \underset{\text{raw score}}{}$$

The score for each problem should reflect how completely the question was answered, that is, the solution that was produced and the steps taken. You should gauge at what point a mistake was made, and determine whether any use of calculus or mathematics was incorrect. Each problem is given a score of between 0 and 9 points. More points should be given for correct answers that include all work in the answer explanation, and less points should be given for incorrect answers and necessary work that was not written down. It might help to have a teacher or an impartial person knowledgeable in calculus decide on the points to be awarded.

THE COMPOSITE SCORE

To obtain your composite score, use the following method:

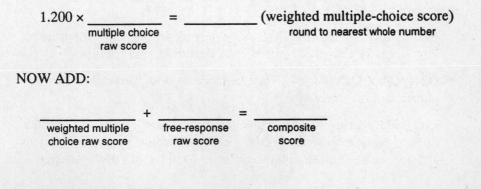

$$1.200 \times \underset{\substack{\text{multiple choice}\\\text{raw score}}}{\underline{\hspace{2cm}}} = \underset{\text{round to nearest whole number}}{\underline{\hspace{2cm}}} \text{ (weighted multiple-choice score)}$$

NOW ADD:

$$\underset{\substack{\text{weighted multiple}\\\text{choice raw score}}}{\underline{\hspace{2cm}}} + \underset{\substack{\text{free-response}\\\text{raw score}}}{\underline{\hspace{2cm}}} = \underset{\substack{\text{composite}\\\text{score}}}{\underline{\hspace{2cm}}}$$

Compare your score with this table to approximate your grade:

AP GRADE	COMPOSITE SCORE
5	78 – 102
4	64 – 77
3	45 – 63
2	30 – 44
1	0 – 29

The overall scores are interpreted as follows: 5-extremely well qualified; 4-well qualified; 3-qualified; 2-possibly qualified; and 1-no recommendation. Most colleges will grant students who earn a 3 or above either college credit or advanced placement. Check with your school guidance office about specific school requirements.

THE ADVANCED PLACEMENT EXAMINATION IN

CALCULUS AB

COURSE REVIEW

CHAPTER 1

ELEMENTARY FUNCTIONS:
Algebraic, Exponential, Logarithmic, and Trigonometric

A. PROPERTIES OF FUNCTIONS

Definition: A function is a correspondence between two sets, the domain and the range, such that for each value in the domain there corresponds exactly one value in the range.

A function has three distinct features:

a) the set x which is the domain,

b) the set y which is the co-domain or range,

c) a functional rule, f, that assigns only one element $y \in Y$ to each $x \in X$. We write $y = f(x)$ to denote the functional value y at x.

Consider Figure 1. The "machine" f transforms the domain X, element by element, into the co-domain Y.

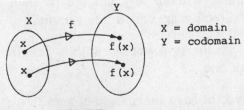

X = domain
Y = codomain

FIGURE 1

■ PARAMETRIC EQUATIONS

If we have an equation $y = f(x)$, and the explicit functional form contains an arbitrary constant called a parameter, then it is called a parametric equation. A function with a parameter represents not one but a family of curves.

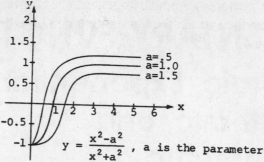

$$y = \frac{x^2 - a^2}{x^2 + a^2} \text{ , a is the parameter}$$

FIGURE 2

Often the equation for a curve is given as two functions of a parameter t, such as

$$X = x(t) \text{ and } Y = y(t).$$

Corresponding values of x and y are calculated by solving for t and substituting.

■ VECTORS

A vector (AB) is denoted $\overrightarrow{AB}$, where B represents the head and A represents the tail. This is illustrated in Figure 3.

The length of a line segment is the magnitude of a vector. If the magnitude and direction of two vectors are the same, then they are equal.

Vectors which can be translated from one position to another without any change in their magnitude or direction are called free vectors.

The unit vector is a vector with a length (magnitude) of one.

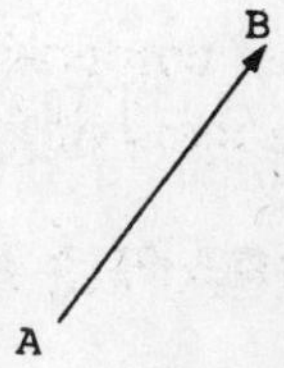

FIGURE 3

The zero vector has a magnitude of zero.

The unit vector, $\vec{i}$, is a vector with magnitude of one in the direction of the x–axis.

The unit vector $\vec{j}$ is a vector with magnitude of one in the direction of the y–axis.

When two vectors are added together, the resultant force of the two vectors produce the same effect as the two combined forces. This is illustrated in Figure 4.

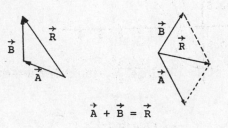

$$\vec{A} + \vec{B} = \vec{R}$$

FIGURE 4

In these diagrams, the vector $\vec{R}$ is called the resultant vector.

■ COMBINATION OF FUNCTIONS

Let f and g represent functions, then

a)   the sum $(f + g)(x) = f(x) + g(x)$,

b)   the difference $(f - g)(x) = f(x) - g(x)$,

c)   the product $(fg)(x) = f(x)\,g(x)$,

d)   the quotient $\left(\dfrac{f}{g}\right)(x) = \dfrac{f(x)}{g(x)}, g(x) \ne 0$,

e)   the composite function $(g \circ f)(x) = g(f(x))$ where $f(x)$ must be in the domain of g.

■ GRAPHS OF A FUNCTION

If (x, y) is a point or ordered pair on the coordinate plane R then x is the first coordinate and y is the second coordinate.

To locate an ordered pair on the coordinate plane simply measure the distance of x units along the x-axis, then measure vertically (parallel to the y-axis) y units.

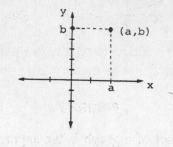

FIGURE 5

This graph illustrates the origin, the x–intercept and the y–intercept.

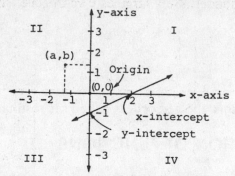

I, II, III, IV are called quadrants in the COORDINATE PLANE.
(*a*, *b*) is an ordered pair with x–coordinate *a* and y–coordinate *b*.

FIGURE 6–Cartesian Coordinate System

The following three graphs illustrate symmetry.

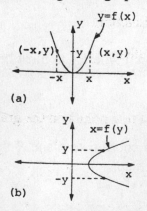

Symmetric about the y–axis

Symmetric about the x–axis
Note: This is not a function of x.

A-4

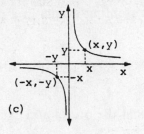

Symmetric about the origin.

(c)

FIGURE 7

Another important part of a graph is the asymptote. An asymptote is a line which will never be touched by the curve as it tends toward infinity.

A vertical asymptote is a vertical line $x = a$, such that the functional value $|f(x)|$ grows indefinitely large as x approaches the fixed value a.

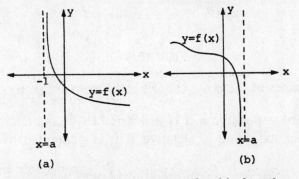

(a)

(b)

$x = a$ is a vertical asymptote for this function

FIGURE 8

The following steps encapsulate the procedure for drawing a graph:

a) Determine the domain and range of the function.

b) Find the intercepts of the graph and plot them.

c) Determine the symmetries of the graph.

d) Locate the vertical asymptotes and plot a few points on the graph near each asymptote.

e) Plot additional points as needed.

■ POLAR COORDINATES

Polar coordinates is a method of representing points in a plane by the use of ordered pairs.

The polar coordinate system consists of an origin (pole), a polar axis and a ray of specific angle.

The polar axis is a line that originates at the origin and extends indefinitely in any given direction.

The position of any point in the plane is determined by its distance from the origin and by the angle that the line makes with the polar axis.

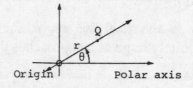

FIGURE 9

The coordinates of the polar coordinate system are (r, θ).

The angle (θ) is positive it if is generated by a counterclockwise rotation of the polar axis, and is negative if it is generated by a clockwise rotation.

The graph of an equation in polar coordinates is a set of all points, each of which has at least one pair of polar coordinates, (r, θ), which satisfies the given equation.

To plot a graph:

1.   Construct a table of values of θ and r.

2.   Plot these points.

3.   Sketch the curve.

■ INVERSE OF A FUNCTION

Assuming that f is a one-to-one function with domain X and range Y, then a function g having domain Y and range X is called the inverse function of f if:

$$f(g(y)) = y \text{ for every } y \in Y \text{ and}$$

$$g(f(x)) = x \text{ for every } x \in X.$$

The inverse of the function f is denoted f^{-1}.

To find the inverse function f^{-1}, you must solve the equation $y = f(x)$ for x in terms of y.

Be careful: This solution must be a function.

■ EVEN AND ODD FUNCTIONS

A function is even if $f(-x) = f(x)$ or

$$f(x) + f(-x) = 2f(x).$$

A function is said to be odd if $f(-x) = -f(x)$ or $f(x) + f(-x) = 0$.

■ ABSOLUTE VALUE

Definition: The absolute value of a real number x is defined as

$$|x| = \begin{cases} x & \text{if } x \geq 0 \\ -x & \text{if } x < 0 \end{cases}$$

For real numbers a and b:

a) $|a| = |-a|$

b) $|ab| = |a| \cdot |b|$

c) $-|a| \leq a \leq |a|$

d) $ab \leq |a||b|$

e) $|a + b|^2 = (a + b)^2$

■ PERIODICITY

A function f with domain X is periodic if there exists a positive real number p such that $f(x + p) = f(x)$ for all $x \in X$.

The smallest number p with this property is called the period of f.

Over any interval of length p, the behavior of a periodic function can be completely described.

■ ZEROES OF A FUNCTION

To locate an ordered pair on the coordinate plane simply measure the distance of x units along the x-axis, then measure vertically (parallel to the y–axis) y units.

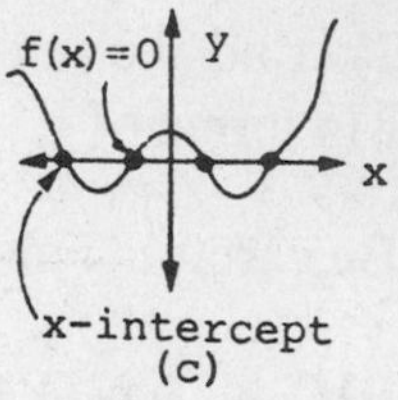

Zeroes of a function
FIGURE 10

B. PROPERTIES OF PARTICULAR FUNCTIONS

In order to graph a trigonometric function, it is necessary to identify the amplitude and the period of the function.

For example, to graph a function of the form

$$y = a \sin (bx + c)$$

a = amplitude   and $^{2\pi}/_b$ = period.

Let us graph the function $y = 2 \sin(2x + {}^\pi/_4)$. Amplitude = 2, period = $^{2\pi}/_2 = \pi$, phase $\} = {}^\pi/_8$.

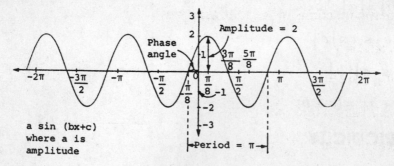

FIGURE 11

The following graphs represent the functions $y = \sin x$ and $y = \cos x$. The amplitude of each is one, while the period of each is 2π.

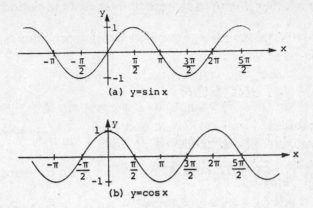

(a) y=sin x

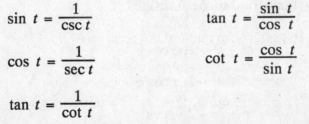

(b) y=cos x

FIGURE 12

■ IDENTITIES AND FORMULAS FOR TRIGONOMETRIC FUNCTIONS

Provided the denominators are not zero, the following relationships exist:

$$\sin t = \frac{1}{\csc t} \qquad\qquad \tan t = \frac{\sin t}{\cos t}$$

$$\cos t = \frac{1}{\sec t} \qquad\qquad \cot t = \frac{\cos t}{\sin t}$$

$$\tan t = \frac{1}{\cot t}$$

If PQR is an angle t and P has coordinates (x, y) on the unit circle, then by joining PR we get angle $PRQ = 90°$ and then we can define all the trigonometric functions in the following way:

sine of t, $\sin t = y$

cosine of t, $\cos t = x$

tangent of t, $\tan t = \frac{y}{x}$, $x \neq 0$

contangent of t, $\tan t = \frac{x}{y}$, $y \neq 0$

secant of t, $\sec t = \frac{1}{x}$, $x \neq 0$

cosecant of t, $\csc t = \frac{1}{y}$, $y \neq 0$.

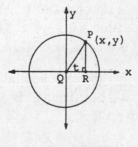

FIGURE 13

$$\cos^2 x + \sin^2 x = 1$$

Therefore,

$$\sec^2 = 1 + \tan^2\theta, \quad \csc^2\theta = 1 + \cot^2\theta$$

$$\sin(A + B) = \sin A \cos B + \cos A \sin B$$

$$\sin(A - B) = \sin A \cos B - \cos A \sin B$$

$$\cos(A + B) = \cos A \cos B - \sin A \sin B$$

$$\cos(A - B) = \cos A \cos B + \sin A \sin B$$

$$\sin^2\theta = \frac{1 - \cos 2\theta}{2}$$

$$\cos^2\theta = \frac{1 + \cos 2\theta}{2}$$

Sine law: $\dfrac{a}{\sin\theta} = \dfrac{b}{\sin\phi} = \dfrac{c}{\sin\psi}$

Cosine law: $a^2 = b^2 + c^2 - 2bc \cos\theta$

$$b^2 = c^2 + a^2 - 2ca \cos\phi$$

$$c^2 = a^2 + b^2 - 2ab \cos\psi$$

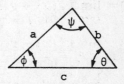

FIGURE 14

■ EXPONENTIAL AND LOGARITHMIC FUNCTIONS

If f is a nonconstant function that is continuous and satisfies the functional equation $f(x + y) = f(x) \cdot f(y)$, then $f(x) = a^x$ for some constant a. That is, f is an exponential function.

Consider the exponential function a^x, $a > 0$ and the logarithmic function $\log_a x$, $a > 0$. Then a^x is defined for all $x \in R$, and $\log_a x$ is defined only for positive $x \in R$.

These functions are inverses of each other,

$$\log_a x = x; \log_a(a^y) = y.$$

A - 10

Let a^x, $a > 0$ be an exponential function. Then for any real numbers x and y

a) $a^x \cdot a^y = a^{x+y}$

b) $(a^x)^y = a^{xy}$

Let $\log_a x$, $a > 0$ be a logarithmic function. Then for any positive real numbers x and y

a) $\log_a(xy) = \log_a(x) + \log_a(y)$

b) $\log_a(x^y) = y \log_a(x)$

C. LIMITS

The following are important properties of limits: Consider

$$\lim_{x \to a} f(x) = L \text{ and } \lim_{x \to a} g(x) = K, \text{ then}$$

A) Uniqueness – If $\lim_{x \to a} f(x)$ exists then it is unique.

B) $\lim_{x \to a} [f(x) + g(x)] = \lim_{x \to a} f(x) + \lim_{x \to a} g(x) = L + K$

C) $\lim_{x \to a} [f(x) - g(x)] = \lim_{x \to a} f(x) - \lim_{x \to a} g(x) = L - K$

D) $\lim_{x \to a} [f(x) \cdot g(x)] = \lim_{x \to a} f(x) \cdot \lim_{x \to a} g(x) = L \cdot K$

E) $\lim_{x \to a} \dfrac{f(x)}{g(x)} = \dfrac{\lim_{x \to a} f(x)}{\lim_{x \to a} g(x)} = \dfrac{L}{K}$ provided $K \neq 0$

■ SPECIAL LIMITS

A) $\lim_{x \to 0} \dfrac{\sin x}{x} = 1,$

B) $\lim_{n \to \infty} \left(1 + \dfrac{1}{n}\right)^n = e,$

Some nonexistent limits which are frequently encountered are:

A) $\lim_{x \to 0} \dfrac{1}{x^2}$, as x approaches zero, x^2 gets very small and also becomes zero therefore $^1/_0$ is undefined and the limit does not exist.

B) $\lim_{x \to 0} \dfrac{|x|}{x}$ does not exist.

■ CONTINUITY

A function f is continuous at a point a if

$$\lim_{x \to a} f(x) = f(a).$$

This implies that three conditions are satisfied:

A) $f(a)$ exists, that is, f is defined at a.

B) $\lim_{x \to a} f(x)$ exists, and

C) the two numbers are equal.

To test continuity at a point $x = a$ we test whether

$$\lim_{x \to a^+} M(x) = \lim_{x \to a^-} M(x) = M(a)$$

■ THEOREMS ON CONTINUITY

A) A function defined in a closed interval $[a, b]$ is continuous in $[a, b]$ if and only if it is continuous in the open interval (a, b), as well as continuous from the right at "a" and from the left at "b."

B) If f and g are continuous functions at a, then so are the functions $f + g$, $f - g$, fg and f/g where $g(a) \neq 0$.

C) If $\lim_{x \to a} g(x) = b$ and f is continuous at b,

$$\lim_{x \to a} f(g(x)) = f(b) = f[\lim_{x \to a} g(x)].$$

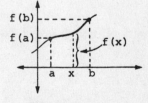

FIGURE 15

D) If g is continuous at a and f is continuous at $b = g(a)$, then
$$\lim_{x \to a} f(g(x)) = f[\lim_{x \to a} g(x)] = f(g(a)).$$

E) Intermediate Value Theorem. If f is continuous on a closed interval $[a, b]$ and if $f(a) \neq f(b)$, then f takes on every value between $f(a)$ and $f(b)$ in the interval $[a, b]$.

CHAPTER 2

DIFFERENTIAL CALCULUS

A. THE DERIVATIVE

■ THE DEFINITION AND Δ-METHOD

The derivative of a function expresses its rate of change with respect to an independent variable. The derivative is also the slope of the tangent line to the curve.

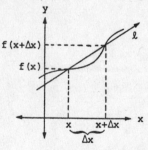

FIGURE 16

Consider the graph of the function f in Figure 16. Choosing a point x and a point $x + \Delta x$ (where Δx denotes a small distance on the x–axis) we can obtain both $f(x)$ and $f(x + \Delta x)$. Drawing a tangent line, l, of the curve through the points $f(x)$ and $f(x + \Delta x)$, we can measure the rate of change of this line. As we let the distance, Δx, approach zero, then

$$\lim_{\Delta x \to 0} \frac{f(x + \Delta x) - f(x)}{\Delta x}$$

becomes the instantaneous rate of change of the function or the derivative.

We denote the derivative of the function f to be f'. So we have

A - 14

$$f'(x) = \lim_{\Delta x \to 0} \frac{f(x + \Delta x) - f(x)}{\Delta x}$$

If $y = f(x)$, some common notations for the derivative are

$$y' = f'(x)$$

$$\frac{dy}{dx} = f'(x)$$

$$D_x y = f'(x) \quad \text{or} \quad Df = f'$$

■ RULES FOR FINDING DERIVATIVES

General Rule:

A) If f is a constant function, $f(x) = c$, then $f'(x) = 0$.

B) If $f(x) = x$, then $f'(x) = 1$.

C) If f is differentiable, then $(cf(x))' = cf'(x)$

D) Power Rule:

If $f(x) = x^n$, $n \in Z$, then

$f'(x) = nx^{n-1}$; if $n < = 0$ then x^n is not defined at $x = 0$.

E) If f and g are differentiable on the interval (a, b) then:

a) $(f + g)'(x) = f'(x) + g'(x)$

b) Product Rule:

$(fg)'(x) = f(x)g'(x) + g(x)f'(x)$

Example:

Find $f'(x)$ if $f(x) = (x^3 + 1)(2x^2 + 8x - 5)$.

$$f'(x) = (x^3 + 1)(4x + 8) + (2x^2 + 8x - 5)(3x^2)$$

$$= 4x^4 + 8x^3 + 4x + 8 + 6x^4 + 24x^3 - 15x^2$$

$$= 10x^4 + 32x^3 - 15x^2 + 4x + 8$$

c) Quotient Rule:

$$\left(\frac{f'}{g}\right)(x) = \frac{g(x)f'(x) - f(x)g'(x)}{[g(x)]^2}$$

Example:

Find $f'(x)$ if $f(x) = \dfrac{3x^2 - x + 2}{4x^2 + 5}$

$$f'(x) = \frac{-(3x^2 - x + 2)(8x) + (4x^2 + 5)(6x - 1)}{(4x^2 + 5)^2}$$

$$= \frac{-(24x^3 - 8x^2 + 16x) + (24x^3 - 4x^2 + 30x - 5)}{(4x^2 + 5)^2}$$

$$= \frac{4x^2 + 14x - 5}{(4x^2 + 5)^2}$$

F) If $f(x) = x^{m/n}$, then

$$f'(x) = \frac{m}{n} x^{\frac{m}{n} - 1}$$

where $m, n \in Z$ and $n \neq 0$.

G) Polynomials. If $f(x) = (a_0 + a_1 x + a_2 x^2 + \ldots + a_n x^n)$ then

$$f'(x) = a_1 + 2a_2 x + 3a_3 x^2 + \ldots + na_n x^{n-1}.$$

This employs the power rule and rules concerning constants.

■ THE CHAIN RULE

Chain Rule: Let $f(u)$ be a composite function, where $u = g(x)$. Then $f'(u) = f'(u)\, g'(x)$ or if $y = f(u)$ and $u = g(x)$ then $D_x y = (D_u y)(D_x u) = f'(u)g'(x)$.

Example:

Find the derivative of: $y = (2x^3 - 5x^2 + 4)^5$.

$$D_x = \frac{d}{dx}.$$

This problem can be solved by simply applying the theorem for $d(u^n)$.

However, to illustrate the use of the chain rule, make the following substitutions:

$$y = u^5 \quad \text{where } u = 2x^3 - 5x^2 + 4$$

Therefore, from the chain rule,

$$D_x y = D_u y \cdot D_x u = 5u^4 (6x^2 - 10x)$$
$$= 5(2x^3 - 5x^2 + 4)^4 (6x^2 - 10x).$$

■ IMPLICIT DIFFERENTIATION

An implicit function of x and y is a function in which one of the variables is not directly expressed in terms of the other. If these variables are not easily or practically separable, we can still differentiate the expression.

Apply the normal rules of differentiation such as the product rule, the power rule, etc. Remember also the chain rule which states

$$\frac{du}{dx} \times \frac{dx}{dt} = \frac{du}{dt}.$$

Once the rules have been properly applied we will be left with, as in the example of x and y, some factors of $\frac{dy}{dx}$.

We can then algebraically solve for the derivative $\frac{dy}{dx}$ and obtain the desired result.

Example:

Find y' in terms of x and y, using implicit differentiation, where

$$y' = \frac{dy}{dx},$$

in the expression:

$$y^3 + 3xy + x^3 - 5 = 0.$$

The derivative of y^3 is $3y^2y'$. The term $3xy$ must be treated as a product. The derivative of $3xy$ is $3xy' + 3y$. The derivative of x^3 is $3x^2$. The derivative of -5 is 0. Therefore,

$$3y^2y' + 3xy' + 3y + 3x^2 = 0.$$

We can now solve for y':

$$y' = -\frac{y + x^2}{y^2 + x}.$$

■ TRIGONOMETRIC DIFFERENTIATION

The three most basic trigonometric derivatives are:

$$\frac{d}{dx}(\sin x) = \cos x,$$

$$\frac{d}{dx}(\cos x) = -\sin x,$$

$$\frac{d}{dx}(\tan x) = \sec^2 x.$$

Given any trigonometric function, it can be differentiated by applying these basics in combination with the general rules for differentiating algebraic expressions.

The following will be most useful if committed to memory:

$D_x \sin u = \cos u \, D_x u$

$D_x \cos u = -\sin u \, D_x u$

$D_x \tan u = \sec^2 u \, D_x u$

$D_x \sec u = \tan u \sec u \, D_x u$

$D_x \cot u = -\csc^2 u \, D_x u$

$D_x \csc u = -\csc u \cot u \, D_x u$

■ INVERSE TRIGONOMETRIC DIFFERENTIATION

Inverse trigonometric functions may be sometimes handled by inverting the expression and applying rules for the direct trigonometric functions.

For example: $y = \sin^{-1} x$

$$D_x y = D_x \sin^{-1} x = \frac{1}{\cos y} = \frac{1}{\sqrt{1 - x^2}}, \, |x| < 1.$$

Here are the derivatives for the inverse trigonometric functions which can be found in a manner similar to the above function:

$$D_x \sin^{-1} u = \frac{1}{\sqrt{1 - u^2}} D_x u, \qquad |u| < 1$$

$$D_x \cos^{-1} u = \frac{-1}{\sqrt{1 - u^2}} D_x u, \qquad |u| < 1$$

$$D_x \tan^{-1} u = \frac{1}{1 + u^2} D_x u, \qquad \text{where } u = f(x) \text{ differentiable}$$

$$D_x \sec^{-1} u = \frac{1}{|u|\sqrt{u^2 - 1}} D_x u, \quad u = f(x), |f(x)| > 1$$

$$D_x \cot^{-1} u = \frac{-1}{1 + u^2} D_x u, \qquad u = f(x) \text{ differentiable}$$

$$D_x \csc^{-1} u = \frac{-1}{|u|\sqrt{u^2 - 1}} D_x u, \quad u = f(x), |f(x)| > 1$$

■ HIGH ORDER DERIVATIVES

The derivative of any function is also a legitimate function which we can differentiate. The second derivative can be obtained by:

$$\frac{d}{dx}\left[\frac{d}{dx} u\right] = \frac{d^2}{dx^2} u = u'' = D^2 u,$$

where $u = g(x)$ is differentiable.

The general formula for higher orders and the nth derivative of u is,

$$\underbrace{\frac{d}{dx} \frac{d}{dx} \cdots \frac{d}{dx}}_{n \text{ times}} u = \frac{d^{(n)}}{dx^n} u = u^{(n)} = D_x^{(n)} u.$$

The rules for first order derivatives apply at each stage of higher order differentiation (e.g., sums, products, chain rule).

A function which satisfies the condition that its nth derivative is zero, is the general polynomial

$$p_{n-1}(x) = a_{n-1} x^{n-1} + a_{n-2} x^{n-2} + \dots + a_0.$$

■ DERIVATIVES OF VECTOR FUNCTIONS

A) Continuity

Let $f(x)$ be a function defined for all values of x near $t = t_0$ as well as at $t = t_0$. Then the function $f(x)$ is said to be continuous at t_0 if

$$\lim_{t \to t_0} f(t) = f(t_0)$$

or, equivalently,

$$\lim_{t \to t_0} f(t) = f(t_0)$$

if and only if for all $\varepsilon > 0$, there exists a $\delta > 0$, such that $|f(t) - f(t_0)| < \varepsilon$, if $|t - t_0| < \delta$.

B) Derivative

The derivative of the vector valued function $V(t)$ with respect to $t \in R$ is defined as the limit

$$\frac{dV(t)}{dt} = \lim_{\Delta t \to 0} \frac{V(t + \Delta t) - V(t)}{\Delta t}.$$

If a vector is expressed in terms of its components along the fixed coordinate axes,

$$V = V_1(t)i + V_2(t)j + V_3(t)k,$$

there follows

$$\frac{dV}{dt} = \frac{dV_1}{dt} i + \frac{dV_2}{dt} j + \frac{dV_3}{dt} k.$$

For the derivative of a product involving two or more vectors the following formulae are used:

$$\frac{d}{dt}(A \cdot B) = A \cdot \frac{dB}{dt} + \frac{dA}{dt} \cdot B$$

$$\frac{d}{dt}(A \times B) = A \times \frac{dB}{dt} + \frac{dA}{dt} \times B$$

$$\frac{d}{dt}(A \cdot B \times C) = \frac{dA}{dt} \cdot (B \times C) + A \cdot \left(\frac{dB}{dt} \times C\right) + A \cdot \left(B \times \frac{dC}{dt}\right).$$

■ PARAMETRIC FORMULA FOR $^{dy}/_{dx}$

According to the chain rule,

$$\frac{dy}{dt} = \frac{dy}{dx} \cdot \frac{dx}{dt}.$$

Since $\frac{dx}{dt} \neq 0$, we can divide through by $\frac{dx}{dt}$ to solve for $\frac{dy}{dx}$. We then obtain the equation

$$\frac{dy}{dx} = \frac{dy}{dt} \div \frac{dx}{dt}.$$

Example:

Find $\dfrac{dy}{dt}$ from

$$y = x^3 - 3x^2 + 5x - 4,$$

where $x = t^2 + t$.

From these equations, we find

$$\frac{dy}{dx} = 3x^2 - 6x + 5$$

$$= 3(t^2 + t)^2 - 6(t^2 + t) + 5,$$

$$\frac{dx}{dt} = 2t + 1.$$

Since

$$\frac{dy}{dt} = \frac{dy}{dx} \frac{dx}{dt}$$

from the chain rule,

$$\frac{dy}{dt} = [3(t^2 + t)^2 - 6(t^2 + t) + 5](2t + 1).$$

We can also first substitute the value of x in terms of t into the equation for y. We then have:

$$y = (t^2 + t)^3 - 3(t^2 + t)^2 + 5(t^2 + t) - 4.$$

When we differentiate this with respect to t, we obtain:

$$\frac{dy}{dt} = 3(t^2 + t)^2 (2t + 1) - 6(t^2 + t)(2t + 1) + 5(2t + 1)$$

$$= [3(t^2 + t)^2 - 6(t^2 + t) + 5](2t + 1),$$

which agrees with the previous answer.

The first method using the chain rule, however, often results in the simpler solution when dealing with problems involving parametric equations.

■ EXPONENTIAL AND LOGARITHMIC DIFFERENTIATION

The exponential function e^x has the simplest of all derivatives. Its derivative is itself.

$$\frac{d}{dx} e^x = e^x \quad \text{and} \quad \frac{d}{dx} e^u = e^u \frac{du}{dx}$$

Since the natural logarithmic function is the inverse of $y = e^x$ and $\ln e = 1$, it follows that

$$\frac{d}{dx} \ln y = \frac{1}{y} \frac{dy}{dx} \quad \text{and} \quad \frac{d}{dx} \ln u = \frac{1}{u} \frac{du}{dx}$$

If x is any real number and a is any positive real number, then

$$a^x = e^{x \ln a}$$

From this definition we obtain the following:

a) $\dfrac{d}{dx} a^x = a^x \ln a \quad \text{and} \quad \dfrac{d}{dx} a^u = a^u \ln a \dfrac{du}{dx}$

b) $\dfrac{d}{dx} (\log_a x) = \dfrac{1}{x \ln a} \quad \text{and} \quad \dfrac{d}{dx} \log_a |u| = \dfrac{1}{u \ln a} \dfrac{du}{dx}$

Sometimes it is useful to take the logs of a function and then differentiate since the computation becomes easier (as in the case of a product).

Steps in Logarithmic Differentiation

1. $y = f(x)$ given

2. $\ln y = \ln f(x)$ take logs and simplify

3. $D_x(\ln y) = D_x(\ln f(x))$ differentiate implicitly

4. $\frac{1}{y} D_x y = D_x(\ln f(x))$

5. $D_x y = f(x) D_x(\ln f(x))$ multiply by $y = f(x)$

To complete the solution it is necessary to differentiate ln $f(x)$. If $f(x) < 0$ for some x then step 2 is invalid and we should replace step 1 by $|y| = |f(x)|$, and then proceed.

Example:

$$y = (x + 5)(x^4 + 1)$$

$$\ln y = \ln[(x + 5)(x^4 + 1)] = \ln(x + 5) + \ln(x^4 + 1)$$

$$\frac{d}{dx}\ln y = \frac{d}{dx}\ln(x + 5) + \frac{d}{dx}\ln(x^4 + 1)$$

$$\frac{1}{y}\frac{dy}{dx} = \frac{1}{x + 5} + \frac{4x^3}{x^4 + 1}$$

$$\frac{dy}{dx} = (x + 5)(x^4 + 1)\left[\frac{1}{x + 5} + \frac{4x^3}{x^4 + 1}\right]$$

$$= (x^4 + 1) + 4x^3(x + 5)$$

This is the same result as obtained by using the product rule.

■ THE MEAN VALUE THEOREM

If f is continuous on $[a, b]$ and has a derivative at every point in the interval (a, b), then there is at least one number c in (a, b) such that

$$f'(c) = \frac{f(b) - f(a)}{b - a}$$

Notice in Figure 17 that the secant has slope

$$\frac{f(b) - f(a)}{b - a}$$

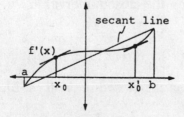

FIGURE 17

and $f'(x)$ has slope of the tangent to the point $(x, f(x))$. For some x_0 in (a, b) these slopes are equal.

Example:

If $f(x) = 3x^2 - x + 1$, find the point x_0 at which $f'(x)$ assumes its mean value in the interval $[2, 4]$.

Recall the mean value theorem. Given a function $f(x)$ which is continuous in $[a, b]$ and differentiable in (a, b), there exists a point x_0 where $a < x_0 < b$ such that:

$$\frac{f(b) - f(a)}{b - a} = f'(x_0),$$

where x_0 is the mean point in the interval.

In our problem, $3x^2 - x + 1$ is continuous, and the derivative exists in the interval $(2, 4)$. We have:

$$\frac{f(4) - f(2)}{4 - 2} = \frac{[3(4)^2 - 4 + 1] - [3(2)^2 - 2 + 1]}{4 - 2}$$

$$= f'(x_0),$$

or

$$\frac{45 - 11}{2} = 17 = f'(x_0) = 6x_0 - 1$$

$$6x_0 = 18$$

$$x_0 = 3.$$

$x_0 = 3$ is the point where $f'(x)$ assumes its mean value.

■ THEOREMS OF DIFFERENTIABLE FUNCTIONS

A) If $f(x)$ is differentiable at x_0, it is continuous there.

B) If $f(x)$ is continuous on the closed interval $[a, b]$, then there is a point $x' \in [a, b]$ for which

$$f(x') < f(x) \ (x : x \in [a, b])$$

C) If $f(x)$ is continuous on the closed interval $[a, b]$, then there is a point x. in $[a, b]$ for which

$$f(x_0) \geq f(x) \ (x : x \in [a, b])$$

D) If $f(x)$ is an increasing function on an interval, then at each point x_0, where $f(x)$ is differentiable we have

$$f'(x_0) \geq 0$$

E) If $f(x)$ is strictly increasing on an interval, and suppose also that $f'(x_0)$ > 0 for some x_0 in the interval, then the inverse function $f^{-1}(x)$ if it exists, is differentiable at the point $y_0 = f(x_0)$.

F) If $f(x)$ is differentiable on the interval $[a, b]$, and $g(x)$ is a differentiable function in the range of f, then the composed function $h = g \circ f$ $(h(x) = g[f(x)])$ is also differentiable on $[a, b]$.

G) Suppose that $f(x)$, $g(x)$ are differentiable on the closed interval $[a, b]$ and that $f'(x) = g'(x)$ for all $x \in [a, b]$, then there is a constant c such that $f(x) = g(x) + C$.

H) Rolle's Theorem

If $f(x)$ is continuous on $[a, b]$, differentiable on (a, b), and $f(a) = f(b) = 0$, then there is a point δ in (a, b) such that $f'(\delta) = 0$.

I) Mean Value Theorem

a) f is continuous on $[a, b]$

b) f is differentiable on (a, b) then there exists some point $\delta \in$ (a, b), such that

$$f'(\delta) = \frac{f(b) - f(a)}{b - a}.$$

■ L'HÔPITAL'S RULE

An application of the Mean Value Theorem is in the evaluation of

$$\lim_{x \to a} \frac{f(x)}{g(x)} \text{ where } f(a) = 0 \text{ and } g(a) = 0.$$

L'Hôpital's Rule states that if the

$$\lim_{x \to a} \frac{f(x)}{g(x)}$$

is an indeterminate form (i.e., $^0/_0$ or $^\infty/_\infty$), then we can differentiate the numerator and the denominator separately and arrive at an expression that has the same limit as the original problem.

Thus,

$$\lim_{x \to a} \frac{f(x)}{g(x)} = \lim_{x \to a} \frac{f'(x)}{g'(x)}$$

In general, if $f(x)$ and $g(x)$ have properties

1) $f(a) = g(a) = 0$

2) $f^{(k)}(a) = g^{(k)}(a) = 0$ for $k = 1, 2, \ldots n;$ but

3) $f^{(n+1)}(a)$ or $g^{(n+1)}(a)$ is not equal to zero, then

$$\lim_{x \to a} \frac{f(x)}{g(x)} = \frac{f^{(n+1)}(x)}{g^{(n+1)}(x)}$$

B. APPLICATION OF THE DERIVATIVE

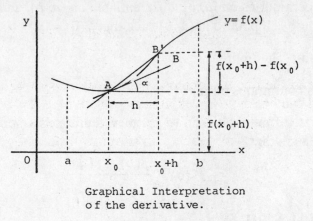

Graphical Interpretation
of the derivative.

FIGURE 18

Graphically the derivative represents the slope of the tangent line AB to the function at the point $(x_0, f(x_0))$.

■ TANGENTS AND NORMALS

Tangents

A line which is tangent to a curve at a point "a", must have the same slope as the curve. That is, the slope of the tangent is simply

$$m = \lim_{h \to 0} \frac{f(a + h) - f(a)}{h}$$

Therefore, if we find the derivative of a curve and evaluate for a specific point, we obtain the slope of the curve and the tangent line to the curve at that point.

A curve is said to have a vertical tangent at a point $(a, f(a))$ if f is continuous at a and

$$\lim_{x \to a} |f'(x)| = \infty.$$

Normals

A line normal to a curve at a point must have a slope perpendicular to the slope of the tangent line. If $f'(x) \neq 0$ then the equation for the normal line at a point (x_0, y_0) is

$$y - y_0 = \frac{-1}{f'(x_0)} (x - x_0).$$

Example:

Find the slope of the tangent line to the ellipse $4x2 + 9y^2 = 40$ at the point $(1, 2)$.

The slope of the line tangent to the curve $4x^2 + 9y^2 = 40$ is the slope of the curve and can be found by taking the derivative, $\frac{dy}{dx}$ of the function and evaluating it at the point $(1, 2)$. We could solve the equation for y and then fine y'. However it is easier to find y' by implicit differentiation. If

$$4x^2 + 9y^2 = 40,$$

then
$$8x + 18y(y') = 0.$$

$$18y(y') = -8x$$

$$y' = \frac{-8x}{18y} = \frac{-4x}{9y}.$$

At the point $(1,2)$, $x = 1$ and $y = 2$. Therefore, substituting these points into $y' = \frac{-4x}{9y}$, we obtain:

$$y' = \frac{-4(1)}{9(2)} = -\frac{2}{9}$$

The slope is $-^2/_9$.

■ MINIMUM AND MAXIMUM VALUES

If a function f is defined on an interval I, then

A) f is increasing on I if $f(x_1) < f(x_2)$ whenever x_1, x_2 are in I and $x_1 < x_2$.

B) f is decreasing on I if $f(x_1) > f(x_2)$ whenever $x_1 < x_2$ in I.

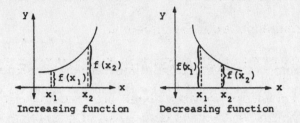

FIGURE 19

C) f is constant if $f(x_1) = f(x_2)$ for every x_1, x_2 in I.

Suppose f is defined on an open interval I and c is a number in I then,

a) $f(c)$ is a local maximum value if $f(x) \le f(c)$ for all x in I.

b) $f(c)$ is a local minimum value of $f(x) \ge f(c)$ for all x in I.

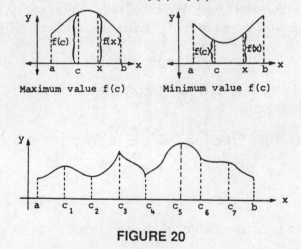

FIGURE 20

Solving Maxima and Minima Problems

Step 1. Determine which variable is to be maximized or minimized (i.e., the dependent variable y).

Step 2. Find the independent variable x.

Step 3. Write an equation involving x and y. All other variables can be eliminated by substitution.

Step 4. Differentiate with respect to the independent variable.

Step 5. Set the derivative equal to zero to obtain critical values.

Step 6. Determine maxima and minima.

■ CURVE SKETCHING AND THE DERIVATIVE TESTS

Using the knowledge we have about local extrema and the following properties of the first and second derivatives of a function, we can gain a better understanding of the graphs (and thereby the nature) of a given function.

A function is said to be smooth on an interval (a, b) if both f' and f'' exist for all $x \in (a, b)$.

The First Derivative Test

Suppose that c is a critical value of a function, f, in an interval (a, b), then if f is continuous and differentiable we can say that,

A) if $f'(x) > 0$ for all $a < x < c$ and $f'(x) < 0$ for all $c < x < b$, then $f(c)$ is a local maximum.

B) if $f'(x) < 0$ for all $a < x < c$ and $f'(x) > 0$ for all $c < x < b$, then $f(c)$ is a local minimum.

C) if $f'(x) > 0$ or if $f'(x) < 0$ for all $x \in (a, b)$ then $f(c)$ is not a local extrema.

Concavity

If a function is differentiable on an open interval containing c, then the graph at this point is

A) concave upward (or convex) if $f''(c) > 0$;

B) concave downward if $f''(c) < 0$.

If a function is concave upward than f' is increasing as x increases. If the function is concave downward, f' is decreasing as x increases.

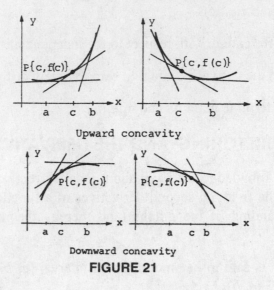

Downward concavity

FIGURE 21

Points of Inflection

Points which satisfy $f''(x) = 0$ may be positions where concavity changes. These points are called the points of inflection. It is the point at which the curve crosses its tangent line.

Graphing a Function Using the Derivative Tests

The following steps will help us gain a rapid understanding of a function's behavior.

A) Look for some basic properties such as oddness, evenness, periodicity, boundedness, etc.

B) Locate all the zeroes by setting $f(x) = 0$.

C) Determine any singularities, $f(x) = \infty$.

D) Set $f'(x)$ equal to zero to find the critical values.

E) Find the points of inflection by setting $f''(x) = 0$.

F) Determine where the curve is concave, $f''(x) < 0$, and where it is convex $f''(x) > 0$.

G) Determine the limiting properties and approximations for large and small $|x|$.

H) Prepare a table of values x, $f(x)$, $f'(x)$ which includes the critical values and the points of inflection.

I) Plot the points found in Step H and draw short tangent lines at each point.

J) Draw the curve making use of the knowledge of concavity and continuity.

Example:

Determine the maxima and minima of $f(x) = x^3 - x$ in the interval from $x = -1$ to $x = 2$.

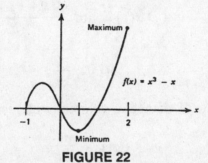

FIGURE 22

To determine the extreme points, we find $f'(x)$, equate it to 0, and solve for x to obtain the critical points. We have:

$$f'(x) = 3x^2 - 1 = 0. \quad x^2 = \frac{1}{3}.$$

Therefore, the critical values are $x = \pm \frac{1}{\sqrt{3}}$. Now

$$f\left(\frac{1}{\sqrt{3}}\right) = -\frac{2}{3\sqrt{3}} \quad \text{and} \quad f\left(-\frac{1}{\sqrt{3}}\right) = \frac{2}{3\sqrt{3}}.$$

Evaluating f at the end points of the interval we have $f(-1) = 0$ and $f(2) = 6$. Therefore, $x = 2$, an end point, is the maximum point for f, and $x = \frac{1}{\sqrt{3}}$ is the minimum point as can be seen in Figure 22. The extreme values of f in $[-1, 2]$ are 6 and $-\frac{2}{3\sqrt{3}}$. The point $\left(-\frac{1}{\sqrt{3}}, \frac{2}{3\sqrt{3}}\right)$ is not an absolute maximum, but it is a relative maximum.

■ RECTILINEAR MOTION

When an object moves along a straight line we call the motion rectilinear motion. Distance s, velocity v, and acceleration a, are the chief concerns of the study of motion.

Velocity is the proportion of distance over time.

$$v = \frac{s}{t}$$

Average velocity $= \dfrac{f(t_2) - f(t_1)}{t_2 - t_1}$

where t_1, t_2 are time instances and $f(t_2) - f(t_1)$ is the displacement of an object.

Instantaneous velocity at time t is defined as

$$v - D\ s(t) = \lim_{h \to 0} \frac{f(t + h) - f(t)}{h}$$

We usually write

$$v(t) = \frac{ds}{dt}.$$

Acceleration, the rate of change of velocity with respect to time is

$$a(t) = \frac{dv}{dt}.$$

It follows clearly that

$$a(t) = v'(t) = s''(t).$$

When motion is due to gravitational effects, $g = 32.2$ ft/sec^2 or $g = 9.81$ m/sec^2 is usually substituted for acceleration.

Speed at time t is defined as $|v(t)|$. The speed indicates how fast an object is moving without specifying the direction of motion.

Example:

A particle moves in a straight line according to the law of motion:

$$s = t^3 - 4t^2 - 3t.$$

When the velocity of the particle is zero, what is its acceleration?

The velocity, v, can be found by differentiating this equation of motion with respect to t. Further differentiation gives the acceleration. Hence, the velocity, v, and acceleration, a, are:

$$v = \frac{ds}{dt} = 3t^2 - 8t - 3,$$

$$a = \frac{dv}{dt} = 6t - 8.$$

The velocity is zero when

$$3t^2 - 8t - 3 = (3t + 1)(t - 3) = 0,$$

from which

$$t = -\frac{1}{3} \text{ or } t = 3.$$

The corresponding values of the acceleration are

$$a = -10 \text{ for } t = -\frac{1}{3}, \text{ and}$$

$$a = +10 \text{ for } t = 3.$$

■ RATE OF CHANGE AND RELATED RATES

Rate of Change

In the last section we saw how functions of time can be expressed as velocity and acceleration. In general, we can speak about the rate of change of any function with respect to an arbitrary parameter (such as time in the previous section).

For linear functions $f(x) = mx + b$, the rate of change is simply the slope m.

For non-linear functions we define the

1) average rate of change between points c and d to be (see Figure 24)

$$\frac{f(d) - f(c)}{d - c}$$

2) instantaneous rate of change of f at the point x to be

$$f'(x) = \lim_{h \to 0} \frac{f(x + h) - f(x)}{h}$$

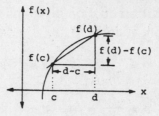

FIGURE 23

If the limit does not exist, then the rate of change of f at x is not defined.

The form, common to all related rate problems, is as follows:

A) Two variables, x and y are given. They are functions of time, but the explicit functions are not given.

B) The variables, x and y are related to each other by some equation such as $x^2 + y^3 - 2x - 7y^2 + 2 = 0$.

C) An equation which involves the rate of change $\frac{dx}{dt}$ and $\frac{dy}{dt}$ is obtained by differentiating with respect to t and using the chain rule.

As an illustration, the previous equation leads to

$$2x\,\frac{dx}{dt} + 3y^2\,\frac{dy}{dt} - 2\,\frac{dx}{dt} - 14y\,\frac{dy}{dt} = 0$$

The derivatives $\frac{dx}{dt}$ and $\frac{dy}{dt}$ in this equation are called the related rates.

EXAMPLE

A point moves on the parabola $6y = x^2$ in such a way that when $x = 6$ the abscissa is increasing at the rate of 2 ft. per second. At what rate is the ordinate increasing at that instant? See Figure 24.

Since

$$6y = x^2,$$

$$6\,\frac{dy}{dt} = 2x\,\frac{dx}{dt}\,,\ or$$

$$\frac{dy}{dt} = \frac{x}{3}\cdot\frac{dx}{dt}. \tag{1}$$

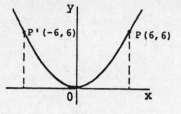

FIGURE 24

This means that, at any point on the parabola, the rate of change of ordinate = $(x/3)$ times the rate of change of abscissa. When $x = 6, \frac{dx}{dt} = 2$ ft. per second. Thus, substitution gives:

$$\frac{dy}{dt} = \frac{6}{3} \cdot 2 = 4 \ ft/\sec.$$

CHAPTER 3

INTEGRAL CALCULUS

A. ANTIDERIVATIVES

Definition:

If $F(x)$ is a function whose derivative $F'(x) = f(x)$, then $F(x)$ is called the antiderivative of $f(x)$.

Theorem:

If $F(x)$ and $G(x)$ are two antiderivatives of $f(x)$, then $F(x) = G(x) + c$, where c is a constant.

Power Rule for Antidifferentiation

Let "a" be any real number, "r", any rational number not equal to -1 and "c" an arbitrary constant.

$$\text{If } f(x) = ax^r, \text{ then } F(x) = \frac{1}{r+1} x^{r+1} + c.$$

Theorem:

An antiderivative of a sum is the sum of the antiderivatives.

$$\frac{d}{dx}(F_1 + F_2) = \frac{d}{dx}(F_1) + \frac{d}{dx}(F_2) = f_1 + f_2$$

B. APPLICATIONS OF ANTIDERIVATIVES $y = y_0 e^{Kt}$: THE LAW OF EXPONENTIAL CHANGE.

Example:

In the course of any given year, the number y of cases of a disease is reduced by 10%. If there are 10,000 cases, today, about how many years will it take to reduce the number of cases to less than 1,000?

$$y = y_0 e^{Kt}$$

$y_0 = 10,000$ so, $y = 10,000\ e^{Kt}$. When $t = 1$, there are 10% fewer cases or 9,000 cases remaining so

$$9,000 = 10,000\ e^{K}$$

$$e^{K} = 0.9 \text{ therefore } K = \ln 0.9$$

Then
$$1,000 = 10,000\ e^{(\ln 0.9)t} \Rightarrow$$

$$.1 = e^{(\ln 0.9)t} \Rightarrow \ln .1 = \ln .9t$$

So
$$t = \frac{\ln 0.1}{\ln 0.9} \approx 21.9 \text{ years}$$

Another application of the antiderivative involves its use with velocity. The following problem illustrates this.

Example:

A body falls under the influence of gravity (gx 32 ft./sec^2) so that its speed is $v = 32t$. Determine the distance it falls in 3 sec. Let x = distance.

$$v = f(t) = 32t$$

The velocity is dependent on time because of the following general relationship:

$$v = gt + v_i$$

where v increases indefinitely as time goes on — neglecting air resistance and some other factors. The initial velocity v_i is zero in this case because the body starts from rest.

Assuming the distance covered is dx in time t, we can represent the velocity in a differential form:

$$\frac{dx}{dt} = v = 32t.$$

Integrating to find the relationship between x and t yields:

$$\int dx = \int 32t \, dt.$$

$$x = \frac{32t^2}{2} + C.$$

$x = 0$ when $t = 0$. Therefore, $0 = 16(0)^2 + C$, $C = 0$.

$$x = 16t^2.$$

The distance the body falls from the reference point,

$$x = 16t^2 = 16(3)^2 = 144 \text{ ft.}$$

C. TECHNIQUES OF INTEGRATION

■ TABLE OF INTEGRALS

$$\int \alpha \, dx = \alpha x + C.$$

$$\int x^n \, dx = \frac{1}{n+1} x^{n+1} + C, \; n \neq 1.$$

$$\int \frac{dx}{x} = \ln |x| + C.$$

$$\int e^x \, dx = e^x + C.$$

$$\int p^x \, dx = \frac{p^x}{\ln p} + C.$$

$$\int \ln x \, dx = x \ln x - x + C.$$

$$\int \cos x \, dx = \sin x + C.$$

$$\int \sin x \, dx = - \cos x + C.$$

$$\int \sec^2 x \, dx = \tan x + C.$$

$$\int \sec x \tan x \, dx = \sec x + C.$$

$\int \tan x \, dx = \ln |\sec x| + C.$

$\int \cot x \, dx = \ln |\sin x| + C.$

$\int \sec x \, dx = \ln |\sec x + \tan x| + C.$

$\int \operatorname{cosec} x \, dx = \ln |\operatorname{cosec} x - \cot x| + C.$

When integrating trigonometric functions, the power rule is often involved. Before applying the fundamental integration formulas, it also may be necessary to simplify the funciton. For that purpose, the common trigonometric identities are most often applicable as, for example, the half-angle formulas and the double-angle formulas. Again, no general rule can be given for finding the solutions. It takes a combination of experience and trial-and-error to learn what to substitute to arrive at the best solution method.

Example:

Integrate:

$$\int \cos x \, e^{2 \sin x} dx.$$

This problem is best solved by the method of substitution. We let $u = 2 \sin x$. Then $du = 2 \cos x \, dx$. Substituting, we obtain:

$$\int \cos x \, e^{2 \sin x} dx = \frac{1}{2} \int e^{2 \sin x} (2 \cos x \, dx)$$

$$= \frac{1}{2} \int e^u \, du = \frac{1}{2} e^u + C$$

$$= \frac{1}{2} e^{2 \sin x} + C.$$

■ INTEGRATION BY PARTS

Differential of a product is represented by the formula

$$d(uv) = udv + vdu$$

Integration of both sides of this equation gives

$$uv = \int u \, dv + \int v \, du \tag{1}$$

or $\qquad \int u \, dv = uv - \int v \, du \tag{2}$

Equation (2) is the formula for integration by parts.

Example:

Evaluate $\int x \ln x \, dx$

Let
$$u = \ln x \quad dv = x dx$$
$$du = 1/x \, dx \quad v = \tfrac{1}{2} x^2$$

Thus,

$$
\begin{aligned}
\int x \ln x \, dx &= (\tfrac{1}{2}) \, x^2 \ln x - \int (\tfrac{1}{2}) x^2 \cdot (\tfrac{1}{x}) \, dx \\
&= (\tfrac{1}{2}) \, x^2 \ln x - \tfrac{1}{2} \int x \, dx \\
&= (\tfrac{1}{2}) \, x^2 \ln x - (\tfrac{1}{4}) x^2 + c
\end{aligned}
$$

Integration by parts may be used to evaluate definite integrals. The formula is:

$$\int_a^b u \, dv = [uv]_a^b - \int_a^b v \, du$$

Example:

Integrate: $\int x \cdot \cos x \cdot dx$.

In this case we use integration by parts, the rule for which states:

$$\int u \, dv = uv - \int v \, du.$$

Let $u = x$ and $dv = \cos x \, dx$. Then $du = dx$ and

$$v = \int \cos x \cdot dx = \sin x.$$

$$\int u \cdot dv = uv - \int v \cdot du$$

becomes
$$\int x \cdot \cos x \cdot dx = x \cdot \sin x - \int \sin x \cdot dx.$$

To integrate $\int \sin x \, dx$ we use the formula, $\int \sin u \, du = -\cos u + C$. This gives:

$$\int x \cdot \cos x \cdot dx = x \sin x - (-\cos x) + C$$

$$= x \sin x + \cos x + C.$$

■ TRIGONOMETRIC SUBSTITUTION

If the integral contains expressions of the form

$$\sqrt{a^2 - x^2} \ , \ \sqrt{a^2 + x^2} \ \text{ or } \ \sqrt{x^2 - a^2} \ ,$$

where $a > 0$, it is possible to transform the integral into another form by means of trigonometric substitution.

General Rules for Trigonometric Substitutions

1. Make appropriate substitutions.

2. Sketch a right triangle.

3. Label the sides of the triangle by using the substituted information.

4. The length of the third side is obtained by use of the Pythagorean Theorem.

5. Utilize sketch, in order to make further substitutions.

  A. If the integral contains the expression of the form $\sqrt{a^2 - x^2}$, make the substitution $x = a \sin \theta$.

$$\sqrt{a^2 - x^2} = \sqrt{a^2 - a^2 \sin^2\theta} = \sqrt{a^2(1 - \sin^2\theta)}$$

$$= \sqrt{a^2 \cos^2\theta} = a \cos \theta.$$

If trigonometric substitution the range of θ is restricted. For example, in the sine substitution the range of $\theta = -\pi/2 \le \theta \le \pi/2$. The sketch of this substitution is shown in Figure 25.

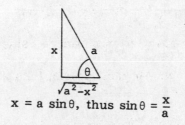

$$x = a \sin\theta, \text{ thus } \sin\theta = \frac{x}{a}$$

FIGURE 25

B. If the integral contains the expression of the form $\sqrt{x^2 - a^2}$, make the substitution $x = a \sec \theta$. The sketch is shown in Figure 26.

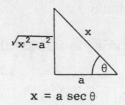

$$x = a \sec \theta$$

FIGURE 26

C. If the integral contains the expression of the form $\sqrt{a^2 + x^2}$, make the substitution $x = a \tan \theta$. The sketch is shown in Figure 27.

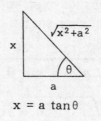

$$x = a \tan \theta$$

FIGURE 27

Example:

Evaluate $\int \dfrac{dx}{\sqrt{4 + x^2}}$

Let $x = 2 \tan \theta$; $dx = 2 \sec^2 \theta\, d\theta$

Thus, $\displaystyle \int \frac{dx}{\sqrt{4 + x^2}} = \int \frac{2 \sec^2 \theta\, d\theta}{\sqrt{4 + (2 \tan \theta)^2}}$

$$= \int \frac{2 \sec^2 \theta\, d\theta}{\sqrt{4(1 + \tan^2 \theta)}}$$

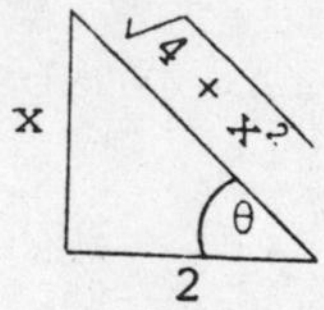

FIGURE 28

A - 42

$$= \int \frac{2\sec^2\theta \, d\theta}{2\sqrt{\sec^2\theta}} = \int \sec\theta \, d\theta$$

$$= \ln|\sec\theta + \tan\theta| + c$$

To convert from θ back to x we use Figure 28 to find:

$$\sec\theta = \frac{\sqrt{4+x^2}}{2} \quad \text{and} \quad \tan\theta = \frac{x}{2}$$

Therefore,

$$\int \frac{dx}{\sqrt{4+x^2}} = \ln\left|\frac{\sqrt{4+x^2}}{2} + \frac{x}{2}\right| + c.$$

Summary of Trigonometric Substitutions

Given expression	Trigonometric substitution
$\sqrt{x^2 - a^2}$	$x = a\sec\theta$
$\sqrt{x^2 + a^2}$	$x = a\tan\theta$
$\sqrt{a^2 - x^2}$	$x = a\sin\theta$

D. THE DEFINITE INTEGRAL

■ AREA

To find the area under the graph of a function f from a to b, we divide the interval $[a, b]$ into n subintervals, all having the same length $(b - a)/n$. This is illustrated in the following figure.

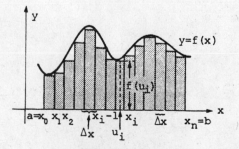

FIGURE 29

Since f is continuous on each subinterval, f takes on a minimum value at some number u_i in each subinterval.

We can construct a rectangle with one side of length $[x_{i-1}, x_i]$, and the other side of length equal to the minimum distance $f(u_i)$ from the x–axis to the graph of f.

The area of this rectangle is $f(u_i)$ Δx. The boundary of the region formed by the sum of these rectangles is called the inscribed rectangular polygon.

The area (A) under the graph of f from a to b is

$$A = \lim_{\Delta x \to 0} \sum_{i=1} f(u_i)\Delta x.$$

The area A under the graph may also be obtained by means of circumscribed rectangular polygons.

In the case of the circumscribed rectangular polygons the maximum value of f on the interval $[x_{i-1}, x_i]$, v_i, is used.

Note that the area obtained using circumscribed rectangular polygons should always be larger than that obtained using inscribed rectangular polygons.

■ DEFINITION OF DEFINITE INTEGRAL

Definition:

Let f be a function that is defined on a closed interval $[a, b]$ and let P be a partition of $[a, b]$. A Riemann Sum of f for P is any expression R_p of the form,

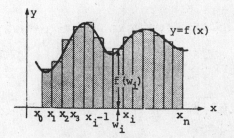

FIGURE 30

A - 44

$$R_p = \sum_{i=1}^{n} f(w_i)\Delta x_i,$$

where w_i is some number in $[x_{i-1}, x_i]$, for $i = 1, 2, ..., n$.

Definition:

Let f be a function that is defined on a closed interval $[a, b]$. The definite integral of f from a to b, denoted by

$$\int_a^b f(x)\ dx$$

is given by

$$\int_a^b f(x)\ dx = \lim_{P \to 0} \sum_i f(w_i)\ \Delta x_i,$$

provided the limit exists.

Theorem:

If f is continuous on $[a, b]$, then f is integrable on $[a, b]$.

Theorem:

If $f(a)$ exists, then

$$\int_a^a f(x)\ dx = 0.$$

■ PROPERTIES OF THE DEFINITE INTEGRAL

A) If f is integrable on $[a, b]$, and k is any real number, then kf is integrable on $[a, b]$ and

$$\int_a^b kf(x)\ dx = k \int_a^b f(x)\ dx.$$

B) If f and g are integrable on $[a,b]$, then $f + g$ is integrable on $[a, b\}$ and

$$\int_a^b [f(x) + g(x)]\ dx = \int_a^b f(x)\ dx + \int_a^b g(x)\ dx.$$

C) If $a < c < b$ and f is integrable on both $[a, c]$ and $[c, b]$, then f is integrable on $[a, b]$ and

$$\int_a^b f(x)\ dx = \int_a^c f(x)\ dx + \int_c^b f(x)\ dx .$$

D) If f is integrable on a closed interval and if a, b, and c are any three numbers in the interval, then

$$\int_a^b f(x)\ dx = \int_a^c f(x)\ dx + \int_c^b f(x)\ dx .$$

E) If f is integrable on $[a, b]$ and if $f(x) \geq 0$ for all x in $[a, b]$, then

$$\int_a^b f(x)\ dx \geq 0 .$$

■ THE FUNDAMENTAL THEOREM OF CALCULUS

The fundamental theorem of calculus establishes the relationship between the indefinite integrals and differentiation by use of the mean value theorem.

Mean Value Theorem for Integrals

If f is continuous on a closed interval $[a, b]$, then there is some number P in the open interval (a, b) such that

$$\int_a^b f(x)\ dx = f(P)\ (b - a)$$

To find $f(P)$ we divide both sides of the equation by $(b - a)$ obtaining

$$f(P) = \frac{1}{b - a} \int_a^b f(x)\ dx .$$

Definition of the Fundamental Theorem

Suppose f is continuous on a closed interval $[a, b]$, then

a) If the function G is defined by:

$$G(x) = \int_a^x f(t)\ dt ,$$

for all x in $[a, b]$, then G is an antiderivative of f on $[a, b]$.

b) If F is any antiderivative of f, then

$$\int_a^b f(x)\ dx = F(b) - F(a)$$

E. APPLICATIONS OF THE INTEGRAL

■ AREA

If f and g are two continuous functions on the closed interval $[a, b]$, then the area of the region bounded by the graphs of these two functions and the ordinates $x = a$ and $x = b$ is

$$A = \int_a^b [f(x) - g(x)]\, dx.$$

where
$$f(x) \geq 0 \quad \text{and} \quad f(x) \geq g(x)$$

$$a \leq x \leq b$$

This formula applies whether the curves are above or below the x–axis.

The area below $f(x)$ and above the x–axis is represented by

$$\int_a^b f(x)$$

The area between $g(x)$ and the x–axis is represented by $\int g(x)$.

Example:

Find the area of the region bounded by the curves
$$y = x^2 \quad \text{and} \quad y = \sqrt{x}.$$

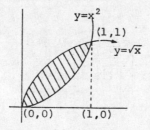

FIGURE 31

$$\text{Area} = A = \int_0^1 (\sqrt{x} - x^2)\, dx$$

$$= \int_0^1 \sqrt{x}\, dx = \int_0^1 x^2\, dx$$

$$= \left[\frac{2}{3} x^{\frac{3}{2}} - \frac{1}{3} x^3 \right]_0^1$$

$$A = \left[\frac{2}{3} - \frac{1}{3} \right] = \frac{1}{3}$$

■ VOLUME OF A SOLID OF REVOLUTION

If a region is revolved about a line, a solid called a solid of revolution is formed. The solid is generated by the region. The axis of revolution is the line about which the revolution takes place.

There are several methods by which we may obtain the volume of a solid of revolution. We shall now discuss three such methods.

1. Disk Method

The volume of a solid generated by the revolution of a region about the x–axis is given by the formula

$$V = \pi \int_a^b [f(x)]^2 \, dx,$$

provided that f is a continuous, nonnegative function on the interval [a, b].

2. Shell Method

This method applies to cylindrical shells exemplified by

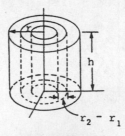

FIGURE 32

The volume of a cylindrical shell is

$$V = \pi r_2^2 h - \pi r_1^2 h$$

$$= \pi (r_2 + r_1)(r_2 - r_1) h$$

$$= 2\pi\left(\frac{r_2 + r_1}{2}\right)(r_2 - r_1)\,h$$

where r_1 = inner radius

r_2 = outer radius

h = height.

Let $r = \frac{r_1 + r_2}{2}$ and $\Delta r = r_2 - r_1$, then the volume of the shell becomes

$$V = 2\pi\, rh\Delta r$$

The thickness of the shell is represented by Δr and the average radius of the shell by r.

Thus,

$$V = 2\pi\int_a^b xf(x)\,dx$$

is the volume of a solid generated by revolving a region about the y-axis. This is illustrated by Figure 33.

3. Parallel Cross Sections

A cross section of a solid is a region formed by the intersection of a solid by a plane. This is illustrated by Figure 34.

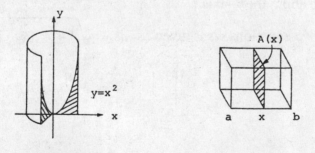

FIGURE 33   **FIGURE 34**

If x is a continuous function on the interval $[a, b]$, then the volume of the cross sectional area $A(x)$ is

$$V = \int_a^b A(x)\,dx.$$

■ AREA OF SURFACE OF REVOLUTION

A surface of revolution is generated when a plane is revolved about a line.

If f' and g' are two continuous functions on the interval $[a, b]$ where $g(t) = 0$, $x = f(t)$ and $y = g(t)$ then, the surface area of a plane revolved about the x–axis is given by the formula

$$S = \int_a^b 2\pi \ g(t) \ \sqrt{[f'(t)]^2 + [g'(t)]^2} \ dt$$

Since $x = f(t)$ and $y = g(t)$,

$$S = \int_a^b 2\pi \ y \ \sqrt{\left(\frac{dx}{dt}\right)^2 + \left(\frac{dy}{dt}\right)^2} \ dt$$

If the plane is revolved about the y–axis, then the surface area is

$$S = \int_a^b 2\pi \ x \ \sqrt{\left(\frac{dx}{dt}\right)^2 + \left(\frac{dy}{dt}\right)^2} \ dt$$

These formulas can be simplified to give the following:

$$S = 2\pi y \int_a^b ds$$

for revolution about the x–axis, and

$$S = 2\pi x \int_a^b ds$$

for revolution about the y–axis.

In the above equations, ds is given as $ds = \sqrt{1 + f'(x)^2} \ dx$.

THE ADVANCED PLACEMENT EXAMINATION IN

CALCULUS AB

TEST I

ADVANCED PLACEMENT CALCULUS AB EXAM I

SECTION I

PART A

Time: 45 minutes
        25 questions

DIRECTIONS: Each of the following problems is followed by five choices. Solve each problem, select the best choice, and blacken the correct space on your answer sheet. Calculators may not be used for this section of the exam.

NOTE:
   Unless otherwise specified, the domain of function f is assumed to be the set of all real numbers x for which $f(x)$ is a real number.

1. $\int_{-2}^{-1} \sqrt{2}\ x^{-2}\ dx$ is approximately

   (A) –0.707   (D) –2.475

   (B) 0.619   (E) 0.707

   (C) 2.475

2.   If $f(x) = \pi^2$, then $f'(1) =$

   (A) 2π (D) 1

   (B) 0   (E) π^2

   (C) π

3.   If $y = \dfrac{1}{\sqrt[3]{e^x}}$, then $y'(1)$ is approximately:

(A)   0.239   (D)   −0.088

(B)   2.150   (E)   0.171

(C)   −0.239

4. $\displaystyle \lim_{h \to 0} \frac{\sin(\pi + h) - \sin \pi}{h}$

(A) 1   (D) $+\infty$

(B) 0   (E) $-\infty$

(C) −1

5.   The slope of the line tangent to the curve $y^3 + x^2y^2 - 3x^3 = 9$ at (1.5, 2) is approximately:

(A)   0.458   (D)   −2.290

(B)   3.206   (E)   −3.206

(C)   −11.450

4

6. If $f'(x) = \sin x$ and $f(\pi) = 3$, then $f(x) =$

   (A) $\cos x + 4$ (D) $\cos x + 3$

   (B) $-\cos x + 2$ (E) $-\cos x - 2$

   (C) $-\cos x + 4$

7. The position of a particle moving along a straight line at any time t is given by $s(t) = 2t^3 - 4t^2 + 2t - 1$.
   What is the acceleration of the particle when $t = 2$?

   (A) 32   (D) 8

   (B) 16   (E) 0

   (C) 4

8. If $f[g(x)] = \sec(x^3 + 4)$, $f(x) = \sec x^3$, and $g(x)$ is <u>not</u> an integer multiple of $\frac{\pi}{2}$, then $g(x)$

   (A) $\sqrt[3]{x} + 4$ (D) $\sqrt[3]{x} - 4$

   (B) $\sqrt[3]{x} - 4$ (E) $\sqrt[3]{x} + 4$

   (C) $\sqrt[3]{x^3 + 4}$

5

9. The equation of each horizontal asymptote for $f(x) = \dfrac{1 - |x|}{x}$ is

(A) $y = 1$ (D) $y = 0$

(B) $y = -1$ (E) $y = 1,\ y = -1$

(C) $x = 0,\ x = 1,\ x = -1$

10. The acceleration of a particle moving on a line is $a = t^{-\frac{1}{2}} + 3t^{\frac{1}{2}}$. What approximate velocity did the particle have from $t = 0$ to $t = 9.61$?

(A) 65.782   (D) 1

(B) 68.782   (E) 45.782

(C) -1

11. The domain of the function defined by $f(x) = \ln(x^2 - x - 6)$ is the set of all real numbers x such that

(A) $x > 0$ (D) $-2 < x < 3$

(B) $-2 \le x \le 3$ (E) $-2 > x$ or $x > 3$

(C) $-2 \le x$ or $x \ge 3$

12. $\int_{1}^{\sqrt{5}} \frac{\ln(x^2)}{x} dx$ is approximately:

(A)   1.296

(D)   0.648

(B)   0.420

(E)   0.805

(C)   1

13. If $y = \arccos(\cos^4 x - \sin^4 x)$, then $y'' =$

(A) 2

(D) $-2(\sin x + \cos x)$

(B) 0

(E) -1

(C) $-2(\cos x - \sin x)$

14. If $\frac{f(x_1)}{f(x_2)} = f\left(\frac{x_1}{x_2}\right)$ for all real numbers x_1 and x_2, where $x_2 \neq 0$ and $f(x_2) \neq 0$, which of the following could define f?

(A) $f(x) = \frac{1}{x}$

(D) $f(x) = \ln x$

(B) $f(x) = x^2 + 3$

(E) $f(x) = e^x$

(C) $f(x) = x + 1$

7

15. $\dfrac{\ln(x^3 e^x)}{x} =$

(A) $\dfrac{3(\ln x + e^x)}{x}$

(D) $\dfrac{3\ln x + x}{x}$

(B) $\ln(x^3 e^x - x)$

(E) $\dfrac{3\ln x}{x}$

(C) $\ln x^2 + 1$

16. $\displaystyle\lim_{x \to 1} \dfrac{\dfrac{1}{x+1} - \dfrac{1}{2}}{x-1} =$

(A) $-\dfrac{1}{4}$

(D) 0

(B) -1

(E) does not exist

(C) $\dfrac{1}{4}$

17. If $\dfrac{r^2}{r-1} \geq r$, then

(A) $r \geq 0$

(D) $r \leq 0$ or $r \geq 1$

(B) $r \leq 0$

(E) $0 \leq r < 1$

(C) $r \leq 0$ or $r > 1$

8

18. If $f'(c) = 0$ for $f(x) = 3x^2 - 12x + 9$,
where $0 \leq x \leq 4$, then $c =$

(A) 2   (D) 1

(B) 3   (E) $\frac{1}{3}$

(C) 0

19. $\lim\limits_{x \to 9} \dfrac{x - 9}{3 - \sqrt{x}} =$

(A) 6   (D) -12

(B) -6 (E) $+\infty$

(C) 0

20. $\int \left(x - \frac{1}{x}\right)^2 dx =$

(A) $\frac{1}{3}\left(x - \frac{1}{x}\right)^3 + C$

(B) $\frac{1}{3}\left(x - \frac{1}{x}\right)^3 \left(1 + \frac{1}{x^2}\right) + C$

(C) $\frac{1}{3}x^3 - 2x - \frac{1}{x^2} + C$

(D) $\frac{1}{3}x^3 - 2x - \frac{1}{x} + C$

(E) $\frac{1}{3}(1 - \ln x)^3 + C$

21. If $e^{g(x)} = \dfrac{x^x}{x^2 - 1}$, then $g(x) =$

(A) $x \ln x - 2x$

(D) $\dfrac{x \ln x}{\ln(x^2 - 1)}$

(B) $\dfrac{\ln x}{2}$

(E) $x \ln x - \ln(x^2 - 1)$

(C) $(x - 2) \ln x$

22. If $h(x) = \dfrac{x^2 + 1}{x^2}$ where $x > 1$, then $h^{-1}(x) =$

(A) $\dfrac{1}{\sqrt{x - 1}}$

(D) $\dfrac{1}{\sqrt{x - 1} + 1}$

(B) $\sqrt{\dfrac{x}{1 + 2x}}$

(E) $\dfrac{1}{-\sqrt{x - 1}}$

(C) $\dfrac{-1}{\sqrt{x}}$

23. If $f(x) = \begin{cases} \dfrac{2x - 6}{x - 3} & x \neq 3 \\ 5 & x = 3 \end{cases}$, then $\lim\limits_{x \to 3} f(x) =$

(A) 5

(D) 6

(B) 1

(E) 0

(C) 2

24. If $f(x) = \dfrac{\sqrt{x+2}}{x+2}$ and $g(x) = \dfrac{1}{x} - 2$, then $f[g(x)] =$

(A) $\dfrac{\sqrt{\dfrac{1}{x} - 2}}{\dfrac{1}{x} - 2}$

(D) $\sqrt{x}$

(B) $\sqrt{\dfrac{1-2x}{x}}$

(E) $\dfrac{\sqrt{x}}{x}$

(C) $\dfrac{\sqrt{\dfrac{1}{x-2} + 2}}{\dfrac{1}{x-2} + 2}$

25. If $\tan x = 2$, then $\sin 2x =$

(A) $\dfrac{2}{5}$

(D) $\dfrac{4}{3}$

(B) $\dfrac{4\sqrt{5}}{5}$

(E) $\dfrac{2}{3}$

(C) $\dfrac{4}{5}$

Time: 45 minutes
       15 questions

DIRECTIONS: Calculators may be used for this section of the test. Each of the following problems is followed by five choices. Solve each problem, select the best choice, and blacken the correct space on your answer sheet.

NOTES:

1. Unless otherwise specified, answers can be given in unsimplified form.

2. The domain of function f is assumed to be the set of all real numbers x for which $f(x)$ is a real number.

26. If $f(x) = \log_b x$, then $f(bx) =$

  (A) $bf(x)$ (D) $xf(b)$

  (B) $f(b)\, f(x)$ (E) $f(x)$

  (C) $1 + f(x)$

27. If $f(x) = \begin{cases} x + 1 & x \le 1 \\ 3 + ax^2 & x > 1 \end{cases}$, then $f(x)$ is

continuous for $a =$

  (A) 1   (D) 0

  (B) -1 (E) -2

  (C) $\dfrac{1}{2}$

12

28. If $g(x) = \dfrac{-x - f(x)}{f(x)}$, $f(1) = 4$ and $f'(1) = 2$, then

$g'(1) =$

(A) $-\dfrac{1}{2}$ (D) $\dfrac{1}{8}$

(B) $\dfrac{11}{8}$ (E) $-\dfrac{1}{8}$

(C) $\dfrac{3}{16}$

29. The domain of $f(x) = \sqrt{4 - x^2}$ is

(A) $-2 \le x \le 2$ (D) $-2 < x < 2$

(B) $-2 \le x$ or $x \ge 2$ (E) $x \ge 2$

(C) $-2 < x$ or $x > 2$

30. $\displaystyle\int \dfrac{x + e^x}{xe^x}\, dx =$

(A) $-e^{-x} - \dfrac{1}{x^2} + C$ (D) $-\dfrac{1}{e^{2x}} + \ln|x| + C$

(B) $e^{-x} - \ln|x| + C$ (E) $e^{-x} - \dfrac{1}{x^2} + C$

(C) $-e^{-x} - \ln|x| + C$

13

31. The area enclosed by the graphs of $y = x^2$ and $y = 2x + 3$ is:

(A) $\dfrac{38}{3}$

(D) $\dfrac{16}{3}$

(B) $\dfrac{40}{3}$

(E) $\dfrac{32}{3}$

(C) $\dfrac{34}{3}$

32. The volume of revolution formed by rotating the region bounded by $y = x^3$, $y = x$, $x = 0$ and $x = 1$ about the x–axis is represented by

(A) $\pi \displaystyle\int_0^1 (x^3 - x)^2 \, dx$

(D) $\pi \displaystyle\int_0^1 (x^2 - x^6) \, dx$

(B) $\pi \displaystyle\int_0^1 (x^6 - x^2) \, dx$

(E) $2\pi \displaystyle\int_0^1 (x^6 - x^2) \, dx$

(C) $2\pi \displaystyle\int_0^1 (x^2 - x^6) \, dx$

33. The vertical asymptote and horizontal asymptote for

$f(x) = \dfrac{\sqrt{x}}{x+4}$ is

(A) $x = -4$, $y = 0$

(B) no vertical asymptote, $y = 0$

(C) no vertical or horizontal asymptote

(D) $x = -4$, no horizontal asymptote

(E) $x = -4$, $y = 1$

34. If $f(x) = x^3 - x$, then

(A) $\dfrac{\sqrt{3}}{3} = x$ is a local maximum of f

(B) $\dfrac{\sqrt{3}}{3} = x$ is a local minimum of f

(C) $\sqrt{3} = x$ is a local maximum of f

(D) $\sqrt{3} = x$ is a local minimum of f

(E) $-\sqrt{3} = x$ is a local minimum of f

35. If $\displaystyle\int_a^b f(x)\,dx = 0$, then

(A) $f(x) = 0$ (D) $f(-x) = -f(x)$

(B) $a = b$ (E) None of these

(C) $f(x) = 0$ or $a = b$

15

36. Use the calculator to find $\int_{0.1}^{0.2} \sqrt{2} \, X^{-6} \, dx$.

(A) 14

(D) 25

(B) 53

(E) 0.8

(C) 3.5

37. The position of a particle moving along a straight line at any time t is given by $S(t) = 2t^3 - 4t^2 + 2t - 1$. The lowest rate of movement within the time interval $[0, 2]$ is

(A) 4.25

(D) −1.5

(B) 0.5

(E) 3

(C) −0.67

38. The acceleration of a particle moving on a line is

$$a(t) = t^{-\frac{1}{2}} + 3t^{\frac{1}{2}}$$

The distance traveled by the particle from $t = 0$ to $t = 3.61$ is approximately

(A) 632.15

(D) 300.1

(B) 65.78

(E) 78.25

(C) 20.21

16

39. Let $f(x) = 3x^2 - 12x + 7$. If $f(x) = 0$, then x equals

(A) 1 and 2

(D) 0.71 and 3.28

(B) −2.28 and 1

(E) 1 and −6

(C) 3 and 2.5

40. Let $f(x) = x^3 - x$. If $f(-x) = -f(x)$, find x.

(A) − 1 and 1

(D) ±0.58

(B) 0 only

(E) None of these

(C) All x

ADVANCED PLACEMENT CALCULUS AB EXAM I

SECTION II

Time: 1 hour and 30 minutes
        6 problems

DIRECTIONS: Show all your work. Grading is based on the methods used to solve the problem as well as the accuracy of your final answers. Please make sure all procedures are clearly shown.

NOTES:

1. Unless otherwise specified, answers can be given in unspecified form.

2. The domain of function f is assumed to be the set of all real numbers x for which $f(x)$ is a real number.

1.   Let f be the function given by $f(x) = 1 + \dfrac{1}{x} + \dfrac{1}{x^2}$.

   (A)   Find the x and y intercepts.

   (B)   Write an equation for each vertical and each horizontal asymptote for the graph of f.

   (C)   Find the intervals on which f is increasing and decreasing.

   (D)   Find the maximum and minimum value of f.

2.   (A)   Find the slope of the line $2x + y - 7 = 0$.

   (B)   Find the slope of the tangent line to the semicircle $x^2 + y^2 = 5, \; y \geq 0$.

   (C)   Find the point on the semicircle with the tangent line perpendicular to the line $2x + y - 7 = 0$ in part (A).

   (D)   Find the intercepts of the tangent line in part (C).

18

3.   (A) Let f have the properties described below

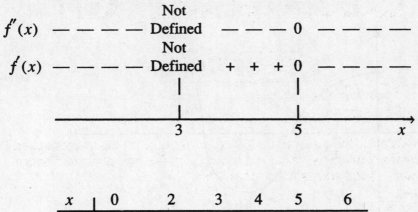

x	0	2	3	4	5	6
$f(x)$	–25	–100	–200	–75	0	–10

$$\lim_{x \to -\infty} f(x) = 0$$

(A) Find the intervals where f is concave down.

(B) Find the equation of each vertical tangent line.

(C) Find each point of inflection of f.

(D) Sketch the graph of f.

4. Let the graph of $s(t)$, the position function (in feet) of a moving particle, be as given below. Let t be time measured in seconds.

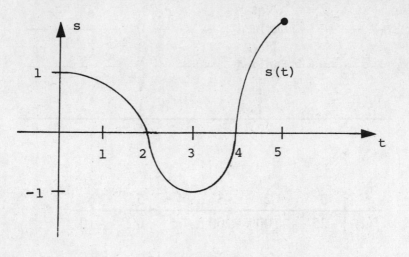

The concavity changes at $t = 2$ and $t = 4$.

(A) Find the values of t for which the particle is moving to the right and when it is moving to the left (i.e., when velocity is positive or negative, respectively).

(B) Find the values of t for which the acceleration is positive and for which it is negative.

(C) Find the values of t for which the particle is speeding up (velocity and acceleration both positive).

5.   Function $f(x) = 2x^2 + x^3$.

   (A)   Find the local maximum of $f(x)$.

   (B)   Find the local minimum of $f(x)$.

   (C)   Evaluate $\int_{-2}^{1} f(x)$.

6.   A flood light is on the ground 45 meters from a building. A thief
      2 meters tall runs from the floodlight directly towards the
      building at 6 meters/sec. How rapidly is the length of his
      shadow on the building changing when he is 15 meters from
      the building?

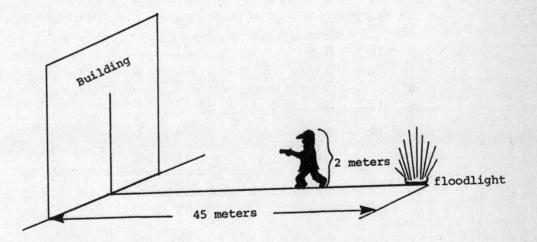

ADVANCED PLACEMENT
CALCULUS AB
EXAM I

ANSWER KEY

Section I

1.	E	21.	E
2.	B	22.	A
3.	C	23.	C
4.	C	24.	D
5.	A	25.	C
6.	B	26.	C
7.	B	27.	B
8.	C	28.	E
9.	E	29.	A
10.	A	30.	C
11.	E	31.	E
12.	D	32.	D
13.	B	33.	B
14.	A	34.	B
15.	D	35.	E
16.	A	36.	B
17.	C	37.	C
18.	A	38.	A
19.	B	39.	D
20.	D	40.	D

Section II

See Detailed Explanations of Answers.

ADVANCED PLACEMENT
CALCULUS AB EXAM I

SECTION I

DETAILED EXPLANATIONS
OF ANSWERS

1. (E)
$$\int_{-2}^{-1} \sqrt{2}\, x^{-2} dx = \sqrt{2} \int_{-2}^{-1} x^{-2} dx$$

$$= \sqrt{2}\left[-x^{-1}\right]_{-2}^{-1}$$

$$= -\sqrt{2}\left[(-1)^{-1} - (-2)^{-1}\right] = \sqrt{2}(-1+2)$$

$$= \frac{\sqrt{2}}{2}$$

$$2\sqrt{x} \div 2 = \approx 0.707$$

2. (B)

$$f(x) = \pi^2 \qquad f'(x) = \frac{d}{dx}(\pi^2) = 0, \text{ since } \pi^2 \text{ is a constant.}$$

3.   (C)

$$y' = \frac{dy}{dx} = \frac{d}{dx}\left(\frac{1}{\sqrt[3]{e^x}}\right) \qquad = \frac{d}{dx}\left(e^{-\frac{x}{3}}\right) = -\frac{1}{3}e^{-\frac{x}{3}}$$

So $y' = -\frac{1}{3}e^{-\frac{x}{3}}$

and $y'(1) = -\frac{1}{3}e^{-\frac{1}{3}} \approx -0.239$

4.   (C)

$$\lim_{h \to 0} \frac{\sin(\pi + h) - \sin\pi}{h} = f'(\pi) \text{ where } f(x) = \sin x.$$

since $f'(x) = \cos x$, $\displaystyle\lim_{h \to 0} \frac{\sin(\pi + h) - \sin\pi}{h}$

$$= \cos\pi$$

$$= -1$$

5.   (A)

$$\frac{d}{dx}\left(y^3 + x^2y^2 - 3x^3\right) = \frac{d}{dx}(9)$$

$$3y^2y' + 2xy^2 + 2yy'x^2 - 9x^2 = 0$$

Note: The product rule must be used when differentiating x^2y^2. Factor y' from the first and third terms.

$$y'\left(3y^2 + 2yx^2\right) = 9x^2 - 2xy^2$$

$$y' = \frac{9x^2 - 2xy^2}{3y^2 + 2yx^2}$$

$\dfrac{dy}{dx}$ at (1.5, 2) is $\dfrac{9(1.5)^2 - 2(1.5)(4)}{3(4) + 2(2)(1.5)}$ $\approx \dfrac{8.25}{18} \approx 0.458$

6.   (B)

$$f(x) = \int f'(x)dx = \int \sin x \, dx = -\cos x + C$$

Since $f(\pi) = 3$ then $-\cos \pi + C = 3$

$$C = 2$$

Therefore $f(x) = -\cos x + 2$.

7.   (B)

$$a(t) = s''(t) = \frac{d}{dt}\left(6t^2 - 8t + 2\right) = 12t - 8$$

$$a''(2) = 24 - 8 = 16$$

8.   (C)

Let $y = g(x)$, then $f[g(x)] = f(y) = \sec y^3$.

Since $f[g(x)] = \sec(x^3 + 4)$ then $\sec(y^3) = \sec(x^3 + 4)$

$$y^3 = x^3 + 4$$
$$y = \sqrt[3]{x^3 + 4}$$

$$f[g(x)] = f(\sqrt[3]{x^3 + 4}) = \sec(\sqrt[3]{x^3 + 4})^3 = \sec(x^3 + 4).$$

Hence $g(x) = \sqrt[3]{x^3 + 4}$

9.   (E)

$$\lim_{x \to +\infty} \frac{1 - |x|}{x} = \lim_{x \to +\infty} \frac{1 - x}{x}$$

$$= \lim_{x \to +\infty} \frac{\frac{1}{x} - 1}{1} = -1$$

25

$$\lim_{x \to -\infty} \frac{1 - |x|}{x} = \lim_{x \to -\infty} \frac{1 + x}{x}$$

$$= \lim_{x \to -\infty} \frac{\frac{1}{x} + 1}{1} = 1$$

The horizontal asymptotes are $y = -1$ and $y = 1$.

10.   (A)

$$\text{velocity} = \int_0^{9.61} \left(t^{-\frac{1}{2}} + 3t^{\frac{1}{2}} \right) dt \qquad = \left(2t^{\frac{1}{2}} + 2t^{\frac{3}{2}} \right) \Big|_0^{9.61}$$

$$= 2(3.1 - 0) + 2\left(3.1^3 - 0 \right)$$

$$= 65.782$$

11.   (E)

The domain is all reals such that $x^2 - x - 6 > 0$

or $(x - 3)(x + 2) > 0$

or $x < -2$ or $x > 3$.

12.   (D)

$$\int_1^{\sqrt{5}} \frac{\ln(x^2)}{x} dx = 2 \int_1^{\sqrt{5}} \frac{\ln x}{x} dx$$

Let $u = \ln x, du = \frac{1}{x} dx$

$u(1) = \ln 1 = 0$

$u(\sqrt{5}) = \ln(\sqrt{5}) = 0.805$

So $= 2 \int_0^{0.805} u \, du = u^2 \Big|_0^{0.805}$

$$= (0.805)^2 - 0 = 0.648$$

13. **(B)**

$$y = \arccos(\cos^4 x - \sin^4 x)$$

$$= \arccos[(\cos^2 x + \sin^2 x)(\cos^2 x - \sin^2 x)]$$

$$= \arccos[(1)(\cos 2x)]$$

$$= 2x$$

$$y' = 2, \quad y'' = 0$$

14. **(A)**

$$\frac{f(x_1)}{f(x_2)} = \frac{\dfrac{1}{x_1}}{\dfrac{1}{x_2}} = \frac{1}{x_1} \cdot \frac{x_2}{1} = \frac{x_2}{x_1} = \frac{\dfrac{1}{x_1}}{x_2} = f\!\left(\frac{x_1}{x_2}\right)$$

15. **(D)**

$$\frac{\ln(x^3 \cdot e^x)}{x} = \frac{\ln x^3 + \ln e^x}{x} = \frac{3\ln x + x}{x}$$

16. **(A)**

$$\lim_{x \to 1} \frac{\dfrac{1}{x+1} - \dfrac{1}{2}}{x - 1}$$

Obtain a common denominator in the main numerator.

$$\lim_{x \to 1} \frac{\dfrac{2 - (x+1)}{2(x+1)}}{x-1} = \lim_{x \to 1} \frac{1-x}{2(x+1)(x-1)}$$

$$= \lim_{x \to 1} \frac{-1}{2(x+1)}$$

$$= \frac{-1}{2(1+1)} = -\frac{1}{4}$$

<u>Note:</u> $\dfrac{1-x}{x-1} = -1$ for $x \neq 1$

27

17.　(C)

$$\frac{r^2}{r-1} \geq r$$

$$\frac{r^2}{r-1} - r \geq 0$$

$$\frac{r^2}{r-1} - r\frac{(r-1)}{r-1} \geq 0 \qquad \text{Express } r \text{ with a denominator of } r-1.$$

$$\frac{r^2 - (r^2 - r)}{r-1} \geq 0$$

$$\frac{r}{r-1} \geq 0$$

$$\frac{r}{r-1} = 0 \qquad \text{when } r = 0 \text{ and undefined when } r = 1.$$

$$\frac{r}{r-1} > 0 \qquad \text{when the numerator and denominator are both positive or both negative}$$

Case (i)　$r > 0$ and $r - 1 > 0$

$r > 0$ and $r > 1$

$\underline{r > 1}$

Case (ii)　$r < 0$ and $r - 1 < 0$

$r < 0$ and $r < 1$

$\underline{r < 0}$

Thus　$\frac{r}{r-1} \geq 0$ when $r \leq 0$ or $r > 1$.

18.　(A)

$$f'(x) = 6x - 12$$

$$f'(c) = 0$$

$$6c - 12 = 0$$

$$c = 2$$

19. (B)

$$\lim_{x \to 9} \frac{x-9}{3-\sqrt{x}}$$ Rationalize the denominator by multiplying by $\dfrac{3+\sqrt{x}}{3+\sqrt{x}}$

$$= \lim_{x \to 9} \frac{(x-9)(3+\sqrt{x})}{(3-\sqrt{x})(3+\sqrt{x})}$$

$$= \lim_{x \to 9} \frac{(x-9)(3+\sqrt{x})}{9-x} \, .$$

Note:

$$\frac{x-9}{9-x} = -1 \text{ for } x \neq 9$$

$$= \lim_{x \to 9} \; -(3+\sqrt{x})$$

$$= -6$$

20. (D)

$$\int \left(x - \frac{1}{x} \right)^2 dx = \int \left(x^2 - 2x \cdot \frac{1}{x} + \frac{1}{x^2} \, dx \right)$$

$$\int (x^2 - 2 + x^{-2}) \, dx$$

$$= \frac{1}{3} x^3 - 2x - \frac{1}{x} + C$$

21. (E)

$$e^{g(x)} = \frac{x^x}{x^2 - 1}$$ Take the natural logarithm of both sides.

$$\ln e^{g(x)} = \ln \left(\frac{x^x}{x^2 - 1} \right)$$

$$g(x) = \ln x^x - \ln(x^2 - 1)$$

$$= x \ln x - \ln(x^2 - 1)$$

22.   (A)

Let $y = \dfrac{x^2 + 1}{x^2}$, $x > 0$; interchange x and y, then solve for y.

$x = \dfrac{y^2 + 1}{y^2}$ multiply by y^2

$xy^2 = y^2 + 1$ subtract y^2 and factor.

$xy^2 - y^2 = 1$

$y^2(x - 1) = 1$ Divide by $x - 1$ ($\neq 0$, since $x > 1$)

$y^2 = \dfrac{1}{x - 1}$

$y = \pm \sqrt{\dfrac{1}{x - 1}}$ Take the square root of both sides

Since $x > 1$ and x was replaced by y, we know $y > 1$, so

$y = + \sqrt{\dfrac{1}{x - 1}} = \dfrac{\sqrt{1}}{\sqrt{x - 1}} = \dfrac{1}{\sqrt{x - 1}}$

Hence

$h^{-1}(x) = \dfrac{1}{\sqrt{x - 1}}$, $x > 1$.

23.   (C)

$\displaystyle \lim_{x \to 3} f(x) = \lim_{x \to 3} \frac{2x - 6}{x - 3} = \lim_{x \to 3} \frac{2(x - 3)}{x - 3} = 2$

24. (D)

$$f[g(x)] = f\left(\tfrac{1}{x} - 2\right) = \frac{\sqrt{\left(\tfrac{1}{x} - 2\right) + 2}}{\left(\tfrac{1}{x} - 2\right) + 2}$$

$$= \frac{\sqrt{\tfrac{1}{x}}}{\tfrac{1}{x}} = \frac{\tfrac{1}{\sqrt{x}}}{\tfrac{1}{x}} = \frac{1}{\sqrt{x}} \cdot \frac{x}{1}$$

Rationalize the
denominator

$$= \frac{x}{\sqrt{x}} \cdot \frac{\sqrt{x}}{\sqrt{x}}$$

$$= \sqrt{x}$$

25. (C)

If $\tan x = 2 = \tfrac{2}{1}$, we have the following diagram. The hypotenuse

is $C = \sqrt{1^2 + 2^2} = \sqrt{5}$.

$$\sin\ 2x = 2 \sin x\ \cos x$$

$$= 2 \left(\frac{2}{\sqrt{5}}\right)\left(\frac{1}{\sqrt{5}}\right)$$

$$= \frac{4}{5}$$

26. (C)

$$f(bx) = \log_b(bx) = \log_b b + \log_b x = 1 + \log_b x$$

$$= 1 + f(x)$$

27. **(B)**

$f(x)$ is continuous for $x < 1$ and $x > 1$ because polynomials are continuous for all reals. We must determine "a" such that $f(x)$ is continuous at $x = 1$.

(i) $f(1) = 1 + 1 = 2$, thus $f(1)$ is defined.

(ii) $\lim\limits_{x \to 1} f(x)$ exists if $\lim\limits_{x \to 1+} f(x) = \lim\limits_{x \to 1-} f(x)$

$\lim\limits_{x \to 1+} f(x) = \lim\limits_{x \to 1+} (3 + ax^2) = 3 + a(1)^2 = 3 + a$

$\lim\limits_{x \to 1-} f(x) = \lim\limits_{x \to 1-} (x + 1) = 1 + 1 = 2$

If $\lim\limits_{x \to 1+} f(x) = \lim\limits_{x \to 1-} f(x)$, then $3 + a = 2$

$$a = -1$$

Hence, $\lim\limits_{x \to 1} f(x) = 2$ if $a = -1$.

(iii)   Since $\lim\limits_{x \to 1} f(x) = f(1)$ then f is continuous when

$a = -1$

The graph of $f(x)$ with $a = -1$ is sketched below:

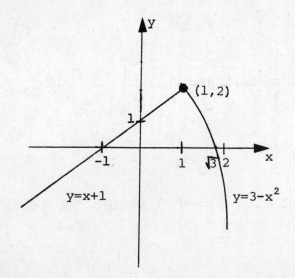

$$\text{(1, 2)}$$
$$y = x + 1 \qquad y = 3 - x^2$$

28. (E)

$$g(x) = -\frac{(x + f(x))}{f(x)} \qquad f(1) = 4 \ , \ f'(1) = 2$$

$$g'(x) = -\left[\frac{(1 + f'(x)) f(x) - f'(x)(x + f(x))}{[f(x)]^2}\right]$$

$$g'(1) = -\left[\frac{(1 + f'(1)) f(1) - f'(1)(1 + f(1))}{(f(1))^2}\right]$$

$$= -\left[\frac{(1 + 2)\,4 - 2\,(1 + 4)}{16}\right]$$

$$= -\left[\frac{12 - 10}{16}\right]$$

$$= -\frac{1}{8}$$

29. (A)

$$4 - x^2 \geq 0$$

$$(2 - x)(2 + x) \geq 0$$

Consider the following sign diagram:

x		−2		2	
$2 - x$	+ + + +	+	0	−	− − −
$2 + x$	− −	0	+ + +	+	+ + +
$(2-x)(2+x)$	− −	0	+ +	0	− − −

Since $(2 - x)(2 + x) \geq 0$ when $-2 \leq x \leq 2$, the domain is $-2 \leq x \leq 2$.

30.   (C)

$$\int \frac{x + e^x}{xe^x}\,dx = \int \left(\frac{x}{xe^x} + \frac{e^x}{xe^x}\right)dx$$

$$= \int \left(e^{-x} + x^{-1}\right)dx$$

$$= -\,e^{-x} + \ln x + C$$

31.   (E)

First determine where the graphs $y = x^2$ and $y = 2x + 3$ intersect.

$$x^2 = 2x + 3$$

$$x^2 - 2x - 3 = 0$$

$$(x - 3)(x + 1) = 0$$

$$x = 3, -1$$

$$A = \int_{-1}^{3} (2x + 3 - x^2)\,dx$$

$$= \left(x^2 + 3x - \frac{1}{3}x^3\right)\Big|_{-1}^{3}$$

$$= 3^2 - (-1)^2 + 3(3 - (-1)) - \frac{1}{3}\left(3^3 - (-1)^3\right)$$

$$= 9 - 1 + 3(4) - \frac{1}{3}(27 + 1)$$

$$= 20 - \frac{28}{3}$$

$$= \frac{32}{3}$$

34

32. (D)

$\pi \int_0^1 x^2 \, dx$ represents the volume of the solid:

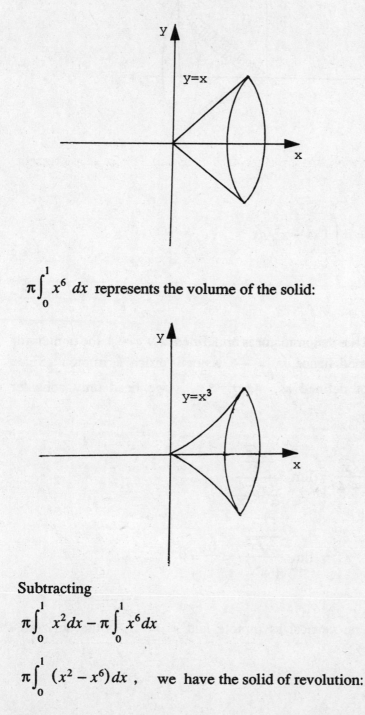

$\pi \int_0^1 x^6 \, dx$ represents the volume of the solid:

Subtracting

$$\pi \int_0^1 x^2 \, dx - \pi \int_0^1 x^6 \, dx$$

$$\pi \int_0^1 (x^2 - x^6) \, dx \, , \quad \text{we have the solid of revolution:}$$

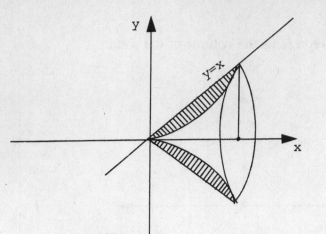

with $V = \pi \int_0^1 (x^2 - x^6)\,dx$

33.　　(B)

Although the denominator is undefined at $x = -4$, the numerator is also undefined, hence, $x = -4$ is not a vertical asymptote. Since $\sqrt{x}$ is not defined as $x \to -\infty$, we need only consider $\lim\limits_{x \to \infty} \dfrac{\sqrt{x}}{x + 4}$.

$$\lim_{x \to \infty} \frac{\sqrt{x}}{x + 4} = \lim_{x \to \infty} \frac{\dfrac{\sqrt{x}}{x}}{\dfrac{x}{x} + \dfrac{4}{x}}$$

$$= \lim_{x \to \infty} \frac{\dfrac{1}{\sqrt{x}}}{1 + \dfrac{4}{x}} = \frac{0}{1} = 0$$

There is no vertical asymptote and $y = 0$ is the horizontal asymptote.

34. **(B)**

$f'(x) = 3x^2 - 1$

$f'(x) = 0$, $3x^2 - 1 = 0$

$$x = \pm \frac{1}{\sqrt{3}} = \pm \frac{\sqrt{3}}{3}$$

$$f\left(\frac{\sqrt{3}}{3}\right) = \left(\frac{\sqrt{3}}{3}\right)^3 - \frac{\sqrt{3}}{3}$$

$$= \frac{\sqrt{3}}{9} - \frac{\sqrt{3}}{3} = \frac{-2\sqrt{3}}{9}$$

$$f\left(\frac{-\sqrt{3}}{3}\right) = -\frac{\sqrt{3}}{9} + \frac{\sqrt{3}}{3} = \frac{2\sqrt{3}}{9}$$

$f''(x) = 6x$

$f''\left(\frac{\sqrt{3}}{3}\right) = 2\sqrt{3} > 0$, $\frac{\sqrt{3}}{3} = x$ is a local minimum

$f''\left(-\frac{\sqrt{3}}{3}\right) = -2\sqrt{3} < 0$, $-\frac{\sqrt{3}}{3} = x$ is a local maxi-

mum (not listed as an answer)

35. **(E)**

$\int_{-1}^{1} x^3 dx = 0$ but $x^3 \neq 0$ and $-1 \neq 1$ which eliminates an-

swers (A), (B) and (C).

$\int_{0}^{\pi} \cos x \, dx = 0$, but $\cos x$ is not an odd function which elimi-

nates answer (D), leaving (E).

36.   (B)

You can solve this problem directly by using

$fnInt\ (\sqrt{2}\ X^\wedge\ (-6),\ x,\ 0.1,\ 0.2)\ .$

Pressing ENTER, 53 will be given.

37.   (C)

The rate of movement of the particle is the velocity $s'(t)$. Use your graphic calculator to draw both $s(t)$ and $s'(t)$.

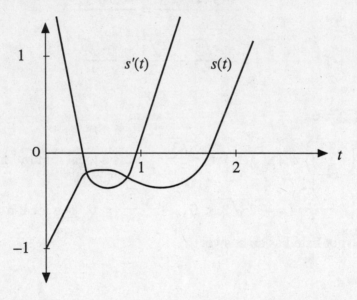

Readjusting the viewing window to $[0, 1] \cdot [-1, 0]$, and tracing $s'(t)$ to the minimum, you can get the lowest rate of movement, -0.67.

38.   (A)

In order to find the distance travelled, you have to know the average velocity of the particle. But velocity is the integral of acceleration, i.e.,

$$\int_0^{9.61} t^{-\frac{1}{2}} + 3t^{\frac{1}{2}}$$

or

$fnInt\ (x^\wedge\ (-0.5) + 3x^\wedge\ 0.5,\ x,\ 0,\ 9.61)$

which gives 65.78.

38

Hence, the distance travelled is

$65.78 \cdot 9.61 = 632.15$.

39.   (D)

Draw the graph of $f(x)$ in the window $[-10, 10] \cdot [-10, 10]$.

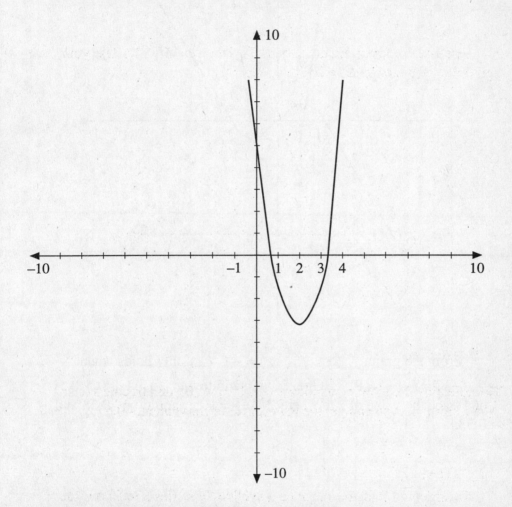

Obviously, when $f(x) = 0$, x can be either between 0 and 1 or between 3 and 4. Change the viewing window to $[0, 4] \cdot [-1, 1]$ and trace the coordinates on the graph. You can find $x = 0.71$, $x = 3.28$.

40.　　(D)

Draw graphs of $f(x)$ and $f'(x)$.

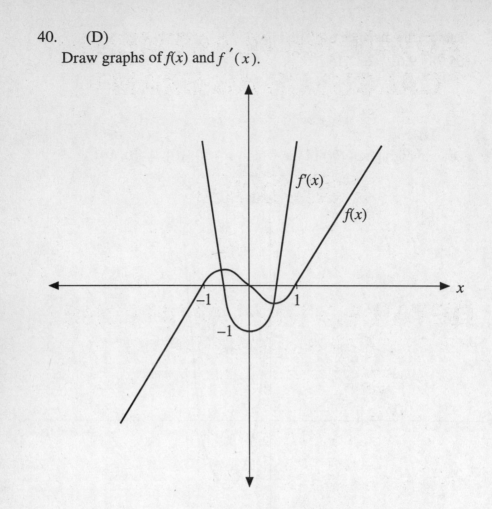

From the graph $f'(x)$, $f'(-x) = -f'(x)$ only holds when $f'(-x) = f'(x) = 0$. Set the viewing window to $[-1, 1] \cdot [-0.5, 0.5]$ and trace the value of x to where $f'(x) = 0$. x should be -0.58 and 0.58.

ADVANCED PLACEMENT CALCULUS AB EXAM I

SECTION II

DETAILED EXPLANATIONS OF ANSWERS

1. (A)

If $x = 0$, $f(x)$ is undefined, so there is no y-intercept.

If $y = f(x) = 0$, then $0 = 1 + \dfrac{1}{x} + \dfrac{1}{x^2}$

$$\Rightarrow \quad 0 = \frac{x^2 + x + 1}{x^2} \quad \Rightarrow x^2 + x + 1 = 0$$

$$\Rightarrow x = \frac{-1 \pm \sqrt{1 - 4}}{2}$$

which gives non-real solutions, so there is no x-intercept.

(B)

$f(x)$ is undefined when $x = 0$ so this is a vertical asymptote. As $x \to \pm\infty$, we see $f(x) \to 1$, so $y = 1$ is a horizontal asymptote.

(C)

$$f'(x) = -\frac{1}{x^2} - \frac{2}{x^3} , \quad f \text{ is increasing when}$$

$$f'(x) = -\frac{1}{x^2} - \frac{2}{x^3} > 0 \quad \Rightarrow \quad -x - 2 > 0 \text{ if } x > 0$$

$$\Rightarrow \quad x < -2 \text{ if } x > 0 \quad ; \text{impossible.}$$

OR $-2 < x$ if $x < 0$, so $-2 < x < 0$ and the interval on which $f(x)$ increases is (-2,0)

$f(x)$ is decreasing if $f'(x) = -\frac{1}{x^2} - \frac{2}{x^3} < 0$.

If $x > 0$, then $-\frac{1}{x^2} - \frac{2}{x^3} < 0$

$$\Rightarrow -x - 2 < 0 \quad \Rightarrow \quad -2 < x \text{ and } x > 0 ,$$

So one interval of decreasing $f(x)$ is $(0,\infty)$

If $x < 0$, then $-\frac{1}{x^2} - \frac{2}{x^3} < 0$

$$\Rightarrow -x - 2 > 0 \quad \Rightarrow \quad -2 > x \text{ , so}$$

another interval on which $f(x)$ is decreasing is $(-\infty, -2)$

(D)
There is no minimum or maximum since $f'(x)$ is never zero, and $f'(x)$ is undefined only at $x = 0$ which is not in the domain of $f(x)$.

2.   (A)

$2x + y - 7 = 0$, $y = 7 - 2x$, slope $= -2$

(B)

By implicit differentiation, $2x + 2yy' = 0$

$$\Rightarrow \; y' = -\frac{2x}{2y} = -\frac{x}{y} = -\frac{x}{\sqrt{5-x^2}}$$

(C)

The slope of the perpendicular must be the negative reciprocal of -2, namely, $1/2$. Now, $-\frac{x}{y} = \frac{1}{2} \; \Rightarrow \; y = -2x$.

Since, $x^2 + y^2 = 5$, we see $x^2 + (-2x)^2 = 5$

$$\Rightarrow x^2 + 4x^2 = 5 \, , \, 5x^2 = 5 \, , \, x = \pm 1$$

If $x = 1$, $y = -2x = -2$

If $x = -1$, $y = -2x = 2$

Since $y \geq 0$ was specified, the point is $(-1, 2)$

(D)

The tangent line has equation $y - 2 = \frac{1}{2}(x + 1)$

$$\Rightarrow \; y = \frac{x}{2} + \frac{5}{2}$$

When $x = 0$, $y = 5/2$ so $(0, 5/2)$ is the y-intercept

When $y = 0$, $x = -5$ so $(-5, 0)$ is the x-intercept.

3.   (A)

$f(x)$ is concave down where $f''(x) < 0$ namely on the intervals $(-\infty, 3)$, $(3, 5)$ and $(5, \infty)$.

(B)

Vertical tangent at $x = 3$.

(C)

No inflection point exists because the concavity (sign of the second derivative) never changes.

(D)

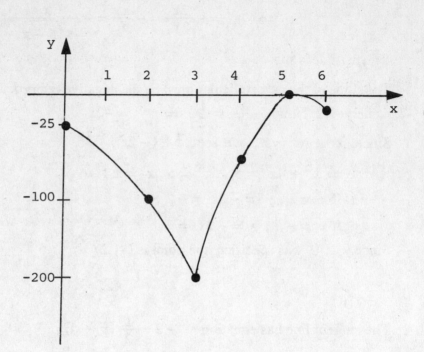

4.   (A)

$s(t)$ is increasing (particle is moving to the right) when $3 < t \leq 5$.

The particle is moving to the left when $s(t)$ is decreasing, namely when $1 \leq t < 3$.

(B)

The acceleration is positive when $s''(t) > 0$, which is when the graph is concave up. This occurs for $2 < t < 4$.

The acceleration is negative when the graph is concave down, namely when $0 \leq t < 2$ or $4 < t < 5$.

(C)

The particle is speeding up when the velocity and acceleration are both positive or both negative. This occurs when $3 \leq t < 4$ or $1 < t < 2$.

5.   **(A)**

Draw the graph of $f(x)$.

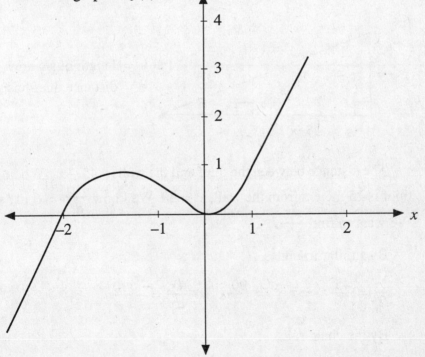

$f(x)$ has a local maximum in $-2 < t < -1$. By using viewing window $[-2,0]$ $[0,2]$, the value of this maximum can be found to be 1.19 at $x=-1.35$.

45

(B)

In the graph, it can be easily seen that $f(x)$ has a local minimum at $x = 0$; its value equals 0.

(C)

By using your calculator, $\int_{-2}^{1} f(x)$ can easily be solved by

$$fnInt(2x \wedge 2 + x \wedge 3, x, -2, 1)$$

which gives 2.25. Hence,

$$\int_{-2}^{1} 2x^2 + x^3 dx = 2.25$$

6.

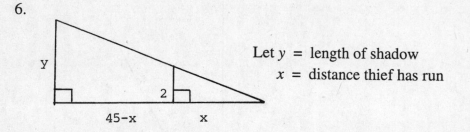

Let y = length of shadow

x = distance thief has run

The distance between the thief and the wall is $45 - x$. When the thief is 15 meters from the wall, $x = 30$. We know $\dfrac{dx}{dt} = 6$ m / sec. We want to find $\dfrac{dy}{dt}$.

By similar triangles,

$$\frac{y}{2} = \frac{45}{x} \implies y = \frac{90}{x}, \text{ so } \frac{dy}{dx} = -\frac{90}{x^2}$$

By the chain rule,

$$\frac{dy}{dt} = \frac{dy}{dx} \frac{dx}{dt}. \text{ At } x = 30, \frac{dy}{dt} = -\frac{90}{30^2} \cdot 6$$

$$= -\frac{540}{900} = -\frac{3}{5} \text{ m / sec.}$$

THE ADVANCED PLACEMENT EXAMINATION IN

CALCULUS AB

TEST II

ADVANCED PLACEMENT CALCULUS AB EXAM II

SECTION I

PART A

Time: 45 minutes
25 questions

DIRECTIONS: Each of the following problems is followed by five choices. Solve each problem, select the best choice, and blacken the correct space on your answer sheet. Calculators may not be used for this section of the exam.

NOTE:
Unless otherwise specified, the domain of function f is assumed to be the set of all real numbers x for which $f(x)$ is a real number.

1. $\displaystyle\int_{1}^{2} \frac{x^3 + 1}{x^2}\ dx =$

(A) 0   (D) 1

(B) $\dfrac{3}{2}$ (E) 3

(C) 2

2.   If $f(x) = \sqrt{1 - x^2}$, which of the following is NOT true?

(A) Domain of $f = [-1, 1]$

(B) $[f(x)]^2 + x^2 = 1$

(C) Range of f is $[0, 1]$

(D) $f(x) = f(-x)$

(E) The line $y = 1$ intersects the graph of f at two points.

49

3. $\lim\limits_{n \to \infty} \left(1 + \frac{1}{n}\right)^{n+2} =$

   (A) e^2 (D) e

   (B) $e + 2$ (E) $e + e^2$

   (C) $2e$

4. $\lim\limits_{x \to 0} \dfrac{\cos^2 x - 1}{2x \sin x} =$

   (A) -1 (D) $\dfrac{1}{2}$

   (B) $-\dfrac{1}{2}$ (E) 0

   (C) 1

5. If $y = \dfrac{1}{\sqrt{2x+3}}$, then $y'(0)$ is approximately:

   (A)   0.193   (D)   5.196

   (B)   −0.096   (E)   −140.296

   (C)   −0.193

6.　If $f(x) = |x|$, then

(A)　Domain of $f' = $ Domain of f.

(B)　$f'(x) = \dfrac{|x|}{x}$　for every real number x.

(C)　$(f'(x))(f(x)) = f(x)$　for every real number x.

(D)　Range of $f' = (-1, 1)$.

(E)　The graph of f' is

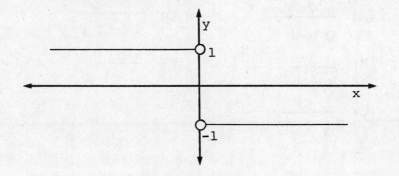

7.　$\displaystyle\int_0^1 (2x+1)^{-3}dx$ is approximately:

(A)　0.888　　　　　　　(D)　−1.500

(B)　−1.111　　　　　　(E)　0.222

(C)　−0.277

8. If $f(x + c) = f(x) \cdot f(c)$ for every real number x and c and $f(0) \neq 1$, then $f(0) =$

(A) 1

(D) –1

(B) 0

(E) $\sqrt{2}$

(C) 0 and 1

9. If $y = \dfrac{x - 1}{x + 1}$, then $\dfrac{dy}{dx} =$

(A) $\dfrac{2x}{(x + 1)^2}$

(D) $-\dfrac{2}{(x + 1)^2}$

(B) $\dfrac{2}{x + 1}$

(E) $\dfrac{2x}{x + 1}$

(C) $\dfrac{2}{(x + 1)^2}$

10. If the graph of f is as in the figure below, where slope of $L_1 = 2$, then $f'(x_0)$ is

(A) $\dfrac{1}{2}$

(D) $-\dfrac{1}{2}$

(B) –2

(E) 0

(C) 2

52

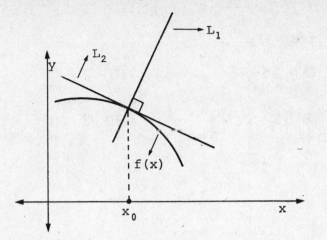

11. If $f(x) = \dfrac{1}{x-2}$ and $\displaystyle\lim_{x \to (-k+1)} f(x)$ does not exist, then
   $k =$

   (A) 2   (D) -2

   (B) 3   (E) -1

   (C) 1

12. If $f(x) = \dfrac{1}{10 - \sqrt{x^2 + 64}}$ is not continuous at c, then $c =$

   (A) 6   (D) ± 4

   (B) -6 (E) 5

   (C) ± 6

13. If $f(x) = \dfrac{1}{x-2}$, $(fg)(1) = 6$, and $g'(1) = -1$,

then $g(1) =$

(A) -5

(D) 7

(B) 5

(E) 8

(C) -7

14. If $f'(x) = \dfrac{\left(x^4 + 5x + 1\right)(12x) - \left(6x^2 - 1\right)\left(4x^3 + 5\right)}{\left(x^4 + 5x + 1\right)^2}$

and $f(0)=2$, then $f(1)$ is approximately:

(A)   0.102

(D)   3.102

(B)   0.714

(E)   0.857

(C)   3.714

15. $\displaystyle\int_0^1 \left(\sec^2 x - \tan^2 x\right)\, dx =$

(A) 3

(D) 4

(B) 5

(E) 1

(C) 2

16. If $2 \leq f(x) \leq (1-x)^2 + 2$ for $x \neq 1$, then $\lim_{x \to 1} f(x) =$

(A) 3

(D) $\frac{5}{2}$

(B) 2

(E) 1

(C) 4

17. If f is differentiable at 0, and $g(x) = [f(x)]^2$, $f(0)$
$= f'(0) = -1$, then $g'(0) =$

(A) -2

(D) 4

(B) -1

(E) 2

(C) 1

18. $\frac{dy}{dx} = x \cot(3x^2)$, then $y =$

(A) $\ln|\sin 3x^2| + c$

(D) $\frac{1}{6}\ln|\sin 3x^2| + c$

(B) $6 \ln|\sin 3x^2| + c$

(E) $\ln c |\sin 3x^2|$

(C) $6 \ln c |\sin 3x^2|$

19. If the k-th derivative of $(3x - 2)^3$ is zero, then k is necessarily

(A) 4

(D) 2

(B) 3

(E) 5

(C) ≥ 4

20. If arc tan $(x) = \ln (y^2)$, then, in terms of x and y, $\dfrac{dy}{dx} =$

(A) $\dfrac{1}{1 - x^2}$

(D) $\dfrac{y}{1 + x^2}$

(B) $\dfrac{-1}{1 - x^2}$

(E) $\dfrac{y}{2(1 + x^2)}$

(C) $\dfrac{y}{1 - x^2}$

21. The volume V(in³) of unmelted ice remaining from a melting ice cube after t seconds is $V = 2000 - 40t + 0.2t^2$. How fast is the volume changing when $t = 40$ seconds?

(A) −26 in³/sec

(D) 0 in³/sec

(B) 24 in³/sec

(E) −24 in³/sec

(C) 120 in³/sec

22. If $\int_a^b f(x)\, dx = 8$, $a = 2$, f is continuous, and the average value of f on [a,b] is 4, then $b =$

(A) 0

(D) 3

(B) 2

(E) 5

(C) 4

23. If $0 \leq x \leq 1$, then $\dfrac{d}{dx} \displaystyle\int_x^0 \dfrac{dt}{2+t} =$

(A) $\dfrac{1}{x+2}$

(D) $\ln|2+x| + c$

(B) $-\dfrac{1}{x+2}$

(E) $-\ln|2+x| + c$

(C) $\ln|2+x|$

24. $\displaystyle\lim_{x \to -1} \dfrac{x+x^2}{x^2-1} =$

(A) $-\dfrac{1}{2}$

(D) $\dfrac{1}{2}$

(B) 1

(E) Does not exist

(C) -1

25. For $x \neq 0$, $\lim_{h \to 0} \frac{1}{h}\left(\frac{1}{x+h} - \frac{1}{x}\right) =$

(A) $\frac{1}{x^2}$

(D) $-\frac{2}{x^2}$

(B) $-\frac{2}{x}$

(E) 0

(C) $-\frac{1}{x^2}$

SECTION I

PART B

Time: 45 minutes
15 questions

DIRECTIONS: Calculators may be used for this section of the test. Each of the following problems is followed by five choices. Solve each problem, select the best choice, and blacken the correct space on your answer sheet.

NOTES:

1. Unless otherwise specified, answers can be given in unsimplified form.

2. The domain of function f is assumed to be the set of all real numbers x for which $f(x)$ is a real number.

26. Let $f(x) = 2\sqrt{x}$. If $f(c) = f'(c)$, then c equals

   (A) 0   (D) 0.49

   (B) 0.82   (E) 2.1

   (C) 1.2

27. A particle moves along the x-axis. Its velocity is given by
$$V(t) = \begin{cases} t^2 & \text{for } 0 \le t \le 2 \\ t+2 & \text{for } t \ge 2 \end{cases}$$
If it starts at the origin, its position after 4 seconds is $x =$

   (A) $37\frac{1}{3}$ (D) $\frac{8}{3}$

   (B) $12\frac{2}{3}$ (E) 6

   (C) 10

28. If $f'(x) = g'(x)$, $f'(x)$ and $g'(x)$ are continuous in $[-1, 1]$, and $f(0) - g(0) = 2$, then

(A) $f(x) - g(x) = -2$

(B) $\int_{-1}^{1} (f(x) - g(x))\, dx = 4$

(C) $\int_{-1}^{1} (f(x) - g(x)) = 0$

(D) The graphs of $f(x)$ and $g(x)$ intersect in $[-1, 1]$.

(E) $\int_{-1}^{1} (g(x) - f(x))\, dx = 4$

29. The area under the first arch $f(x) = \sin x \ln x$ for $x \geq 0$ is

(A) -0.24 (D) -1.75

(B) 0.26 (E) 5.7

(C) 1.5

30. $\lim\limits_{x \to \infty} \dfrac{(1 - 2x^2)^3}{(x^2 + 1)^3}$

(A) 8 (D) ∞

(B) 1 (E) -8

(C) 0

31. The function f defined by $f(x) = x + \frac{1}{x}$ has relative minimum at $x =$

(A) -1

(D) 1

(B) $-\frac{1}{2}$

(E) $\frac{1}{2}$

(C) 0

32. The area of a region bounded by the parabola $8 + 2x - x^2$ and the x-axis is:

(A) $41\frac{1}{3}$

(D) $9\frac{1}{3}$

(B) 36

(E) 24

(C) 20

33. $\int_{-1}^{1} \left| x^2 - 1 \right| \, dx =$

(A) $\frac{4}{3}$

(D) $\frac{2}{3}$

(B) 0

(E) $\frac{5}{3}$

(C) $-\frac{4}{3}$

34. Let $f(x) = x^3 - 2x$. The relationship between its local minimum and local maximum is

(A) $f_{min} = 2f_{max}$

(D) $f_{min} = 1.5f_{max}$

(B) $f_{min} = f_{max}$

(E) $f_{min} = \sqrt{f_{max}}$

(C) $f_{min} = -f_{max}$

35. Which of the following functions is not symmetric with respect to the origin?

(A) $\tan x$

(D) $\sin x$

(B) $\dfrac{1}{x}$

(E) $\cos x$

(C) $\cot x$

36. Estimate the HIGHEST rate of change for the function $y = \sqrt{1-x}$ inside $0 \le x \le 0.8$.

(A) -1.14

(D) -3

(B) 5.00

(E) -2

(C) 0

37.   If $f(x) = x^{\frac{1}{3}} \ln x$, then $f'(2)$ equals

(A) −0.75   (D) 0

(B) 0.95   (E) 0.75

(C) 0.25

38.   Which of the following statements is/are true?

I.   If f is continuous everywhere, then f is differentiable everywhere.

II.   If f is differentiable everywhere, then f is continuous everywhere.

III. If f is continuous and $f(x) \geq 2$ for every x in $[3, 7]$, then

$$\int_{3}^{7} f(x)\, dx > 8.$$

(A) I only   (D) I and III only

(B) II only   (E) II and III only

(C) III only

39. Let $g(x)$ be the inverse of $f(x)$. If $f'(x) = g'(x)$, then $f'(1)$ is necessarily

(A) 1 (D) ± 1

(B) -1 (E) $\dfrac{1}{2}$

(C) 0

40. Which of the following is NOT true about $y = \cos(-x + \pi)$?

(A) y has the same period as $\cos(x - \pi)$.

(B) y has the same period as $\tan\left(2 - \dfrac{x}{2}\right)$.

(C) y has only one inflection point in $(-\pi, \pi)$.

(D) $\dfrac{d^2 y}{dx^2} + y = 0$.

(E) y has minimum at $x = 0$.

ADVANCED PLACEMENT CALCULUS AB EXAM II

SECTION II

Time: 1 hour and 30 minutes
       6 problems

DIRECTIONS: Show all your work. Grading is based on the methods used to solve the problem as well as the accuracy of your final answers. Please make sure all procedures are clearly shown.

NOTES:

1. Unless otherwise specified, answers can be given in unspecified form.

2. The domain of function f is assumed to be the set of all real numbers x for which $f(x)$ is a real number.

1. Show that, if f is continuous and $0 \leq f(x) \leq 1$ for every x in $[0, 1]$, then there exists at least one point c such that $f(c) = c$. (Hint: Apply the intermediate value theorem to $g(x) = x - f(x)$, or try to answer it by sketching graphs which represent the possible cases of the graph of f).

2. Let $f(x) = \ln(x^2 - x - 6)$

    (A) The domain of $f(x)$ is $b < x < a$. Find a and b.

    (B) Find $f(5)$.

    (C) Find $f(-3)$.

3.   If $f(x) = \dfrac{1 - x^2}{x^2 + 1}$, then

(A) Find the domain of f.

(B) Find $\lim\limits_{x \to \infty} f(x)$ and $\lim\limits_{x \to -\infty} f(x)$.

(C) Find the intervals where f increases and where it decreases. Justify your answer.

(D) Find the equation of the tangent line that is parallel to the x - axis.

4.   A conical silver cup 8 inches across the top and 12 inches deep is leaking water at the rate of 2 inches³ per minute. (Figure below.) At what rate is the water level dropping:

(A) when the water is 6 inches deep?

(B) when the cup is half full?

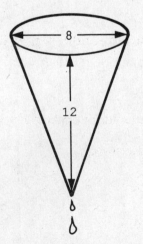

5. $y = f(x)$ is a function where f' and f'' exist and have the following characteristics:

x	$x < -2$	$x = -2$	$-2 < x < 0$	$0 < x < 2$	$x = 2$	$x > 2$
$f'(x)$	+	0	−	−	0	+
$f''(x)$	−	−	−	+	+	+

If $f(-2) = 8$, $f(0) = 4$ and $f(2) = 0$, then

(A) Find all inflection points of $-2f$

(B) Find all relative minimum and relative maximum of $-2f$

(C) Discuss the concavity of $-2f$

(D) Sketch the graph of $-2f$

6. (A) At what values of x do $y = x$ and $y = x^3$ intersect?

(B) Find the area of the region bounded by $y = x$ and $y = x^3$

(C) Find the volume obtained by rotating the region in (B) about the x-axis.

ADVANCED PLACEMENT
CALCULUS AB
EXAM II

ANSWER KEY

Section I

1.	C	21.	E
2.	E	22.	C
3.	D	23.	B
4.	B	24.	D
5.	C	25.	C
6.	D	26.	D
7.	E	27.	B
8.	B	28.	B
9.	C	29.	A
10.	D	30.	E
11.	E	31.	D
12.	C	32.	B
13.	A	33.	A
14.	C	34.	C
15.	E	35.	E
16.	B	36.	A
17.	E	37.	B
18.	D	38.	B
19.	C	39.	D
20.	E	40.	C

Section II

See Detailed Explanations of Answers.

ADVANCED PLACEMENT
CALCULUS AB EXAM II

SECTION I

DETAILED EXPLANATIONS
OF ANSWERS

1. (C)

$$\frac{x^3 + 1}{x^2} = \frac{x^3}{x^2} + \frac{1}{x^2}$$

$$= x + \frac{1}{x^2}$$

Therefore,

$$\int_1^2 \frac{x^3 + 1}{x^2}\, dx = \int_1^2 (x + x^{-2})\, dx$$

$$= \left(\frac{x^2}{2} - \frac{1}{x} \right)\Big|_1^2$$

$$= \left(2 - \frac{1}{2} \right) - \left(\frac{1}{2} - 1 \right)$$

$$= 2$$

Remark: Use parentheses as above to avoid computational errors like: $2 - 1/2 - 1/2 - 1 = 0$.

2. (E)

(i) $y = \sqrt{1 - x^2}$ is defined $\Leftrightarrow 1 - x^2 \geq 0$

$$\Leftrightarrow x^2 \leq 1$$

$$\Leftrightarrow |x| \leq 1$$

Therefore, domain $= [-1, 1]$

(ii) $y \geq 0$ for every x in $[-1, 1]$ and its graph is as follows:

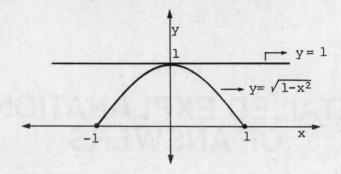

Moreover, its range is $[0, 1]$.

Finally, $f(x)$ is 1 only when $x = 0$. Hence (E) is not true.

3. (D)

$$\left(1 + \frac{1}{n}\right)^{n+2} = \left(1 + \frac{1}{n}\right)^n \cdot \left(1 + \frac{1}{n}\right)^2 \qquad \ldots\ldots(i)$$

Therefore,

$$\lim_{n \to \infty} \left(1 + \frac{1}{n}\right)^{n+2} = \lim_{n \to \infty} \left(1 + \frac{1}{n}\right)^n \cdot \lim_{n \to \infty} \left(1 + \frac{1}{n}\right)^2,$$

from (i). $= e(1)^2$

$= e$

Note: $\left(1 + \frac{1}{n}\right)^{n+2} \neq \left(\left(1 + \frac{1}{n}\right)^n\right)^2$, and therefore, e^2 is incorrect.

70

4. **(B)**

$$\cos^2 x - 1 = -(1 - \cos^2 x) \qquad \ldots\ldots(i)$$

$$= -\sin^2 x$$

Therefore, $\displaystyle\lim_{x\to 0} \frac{\cos^2 x - 1}{2x \sin x} \qquad \ldots\ldots(ii)$

$$= \lim_{x\to 0} \frac{-\sin^2 x}{2x \sin x}, \qquad \text{substituting (i) in (ii)}$$

$$= \lim_{x\to 0} \left(-\frac{1}{2}\right)\frac{\sin x}{x}$$

$$= -\frac{1}{2}, \text{ since } \lim_{x\to 0} \frac{\sin x}{x} = 1$$

You can also use L'Hôpital's rule.

Also see from (i) above that $\cos^2 x - 1 \neq \sin^2 x$. Thus 1/2 is not the limit.

5. **(C)**

$$y = \frac{1}{\sqrt{2x+3}} = (2x+3)^{-\frac{1}{2}}$$

Using chain rule:

$$\frac{dy}{dx} = \frac{1}{2}(2x+3)^{-\frac{1}{2}-1} \cdot \frac{d}{dx}(2x+3)$$

$$= -\frac{1}{2}(2x+3)^{-\frac{3}{2}} \cdot 2$$

$$= -\frac{1}{\sqrt{2x+3}}$$

So $y^1(0) = -\dfrac{1}{\sqrt{(0+3)^3}} = -\dfrac{1}{\sqrt{27}} \approx -0.193$

6.   (D)

$$|x| = \begin{cases} x & \text{for } x > 0 \\ -x & \text{for } x < 0 \end{cases}$$

$$\Rightarrow f'(x) = \begin{cases} 1 & \text{for } x > 0 \\ -1 & \text{for } x < 0 \end{cases} \quad \text{But } f \text{ is not differentiable at } 0.$$

As a result, Domain of $f' = \{x \in R \mid x \neq 0\}$,   while Domain of $f = R$. Consequently, A, B, and C are false. Moreover, the graph of f' is:

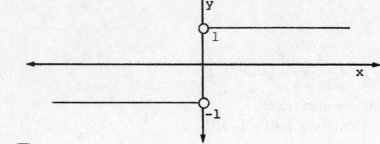

7.   (E)

Let $u = 2x + 1 \Rightarrow \dfrac{du}{dx} = 2$

$\Rightarrow du = 2dx$, by cross multiplication

$\Leftrightarrow \dfrac{1}{2} \bullet du = \dfrac{1}{2} \bullet 2 \bullet dx$, multiplying both sides by $\dfrac{1}{2}$

$\Leftrightarrow dx = \dfrac{du}{2}$

Therefore, $\int_0^1 (2x+1)^{-3}dx = \dfrac{1}{2}\int u^{-3}du \big| u = 2x+1\big|_{x=0}^{x=1}$

$= \dfrac{1}{4}u^{-2}; \quad = \dfrac{-1}{4(2x+1)^2}\Big|_{x=0}^{x=1}; \quad = \left(-\dfrac{1}{4}\cdot\dfrac{1}{9}\right) - \left(-\dfrac{1}{4}\right)$

$= -\dfrac{1}{36} + \dfrac{1}{4}; \quad = \dfrac{2}{9}.$

$2 + 9 = \approx 0.222$

8.   (B)

$f(0 + 0) = f(0) \cdot f(0)$.   On the other hand,

$0 + 0 = 0$

Therefore, $f(0) = [f(0)]^2$

$\Leftrightarrow \ f(0) - [f(0)]^2 = 0$

$\Leftrightarrow \ f(0)\,(1 - f(0)) = 0$

$\Leftrightarrow \ f(0) = 0, \text{ or } f(0) = 1.$

But, $f(0) \neq 1$ (You are given this)

Hence, $f(0) = 0.$

73

9. (C)

$$y = \frac{x - 1}{x + 1}$$

$$\Rightarrow \frac{dy}{dx} = \frac{1 \cdot (x + 1) - (x - 1)}{(x + 1)^2}$$

$$= \frac{x + 1 - x + 1}{(x + 1)^2}$$

$$= \frac{2}{(x + 1)^2} \ .$$

Unlike the product rule, we do not add $(1)(x + 1)$ and $(1)(x - 1)$. If you add them you will get $\frac{2x}{(x + 1)^2}$, which is incorrect.

10. (D)

Referring to the graph in the problem:

Let slope of $L_1 = m_1$

Let slope of $L_2 = m_2$

Therefore, $m_1 m_2 = -1$, since the slopes of two, non-vertical, perpendicular lines are negative reciprocals of each other.

$$\Leftrightarrow m_2 = -\frac{1}{2} \ .$$

Also, L_2 is tangent to the graph of f at x_0.

$$\Rightarrow f'(x_0) = -\frac{1}{2} \ .$$

11. **(E)**

$\dfrac{1}{x-2}$ has no limit only at $x = 2$; see the graph below.

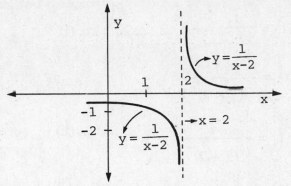

Therefore, $-k + 1 = 2$

$\Leftrightarrow k - 1 = -2$, by multiplying both sides by (-1)

$\Leftrightarrow \quad k = -1$, adding 1 to both sides.

12. **(C)**

$x^2 + 64 \geq 64$ for every real number x.

$10 - \sqrt{x^2 + 64}$ is continuous for every real number x.

But the quotient: $\dfrac{1}{10 - \sqrt{x^2 + 64}}$, is not continuous

only if $10 - \sqrt{x^2 + 64} = 0$

$\Leftrightarrow 10 = \sqrt{x^2 + 64}$

$\Leftrightarrow 100 = x^2 + 64$, squaring both sides

$\Leftrightarrow x^2 = 36$, adding -64 to both sides

$\Leftrightarrow x = \pm 6$.

13. (A)

$(fg)'(1) = f'(1)g(1) + f(1)g'(1)$ by the product rule(*)

$$f(1) = \frac{1}{1-2} = -1 \quad \text{........................ (i)}$$

$$f'(x) = -\frac{1}{(x-2)^2}$$

$$f'(1) = -\frac{1}{(1-2)^2} = -1 \quad \text{............. (ii)}$$

$$g'(1) = -1 \quad \text{.. (iii)}$$

$$(fg)'(1) = 6 \quad \text{....................................... (iv)}$$

Now substitute (i) - (iv) in (*) above:
$$6 = (-1)\,g(1) + (-1)(-1)$$

$$\Leftrightarrow g(1) = -5$$

14. (C)

From the quotient rule the given expression is the derivative of

$$f(x) = \frac{6x^2 - 1}{x^4 + 5x + 1} + c. \text{ But } f(0) = 2$$

$$\Leftrightarrow \frac{6 \bullet 0^2 - 1}{0^4 + 5 \bullet 0 + 1} + c = 2, \quad \text{substituting 0 for } x.$$

$$\Leftrightarrow -1 + c = 2$$
$$\Leftrightarrow c = 3$$

So $f(x) = \dfrac{6x^2 - 1}{x^4 + 5x + 1} + 3.$

And $f(1) = \dfrac{6(1)^2 - 1}{(1)^4 + 5(1) + 1} + 3 = \dfrac{5}{7} + 3 = 3\dfrac{5}{7}$

$$f(1) = 3\frac{5}{7} \approx 3.714$$

15.   (E)

By trigonometric identity: $\sec^2 x = 1 + \tan^2 x$

$$\Rightarrow \sec^2 x - \tan^2 x = 1$$

$$\Rightarrow \int_0^1 (\sec^2 x - \tan^2 x)\, dx$$

$$= \int_0^1 1 \cdot dx$$

$$= 1.$$

<u>Remark</u>: Whenever expressions like
$\sec^2 x - \tan^2 x$, $\cos^2 x - 1$, $\sin^2 x$, etc.
appear in a problem, it is worth trying trigonometric identities first.

16.   (B)

$\lim\limits_{x \to 1} 2 = 2$, and

$\lim\limits_{x \to 1} (1 - x)^2 + 2 = 2$

Therefore, $2 \le \lim\limits_{x \to 1} f(x) \le 2$, from the figure.

$$\Leftrightarrow \lim_{x \to 1} f(x) = 2.$$

77

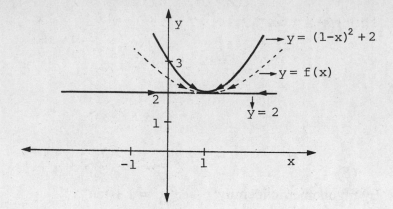

17.   (E)

$$g(x) = [f(x)]^2$$

$$\Rightarrow g'(x) = 2 \cdot [f(x)] \cdot f'(x)$$

Therefore, $g'(0) = 2 \cdot f(0) \cdot f'(0)$

$$= 2 \cdot (-1)(-1)$$

$$= 2$$

A common arithmetic error is: $2(-1)(-1) = -2$.

18.   (D)

By the method of substitution:

Let $u = 3x^2$

$$\Rightarrow \frac{du}{dx} = 6x$$

$$\Leftrightarrow x\,dx = \frac{du}{6} \text{ , after cross multiplication and division.}$$

Therefore, $\int x \cot(3x^2)\,dx = \dfrac{1}{6} \int \cot u\,du$

$$= \frac{1}{6} \int \frac{\cos u}{\sin u}\,du$$

$$= \frac{1}{6} \ln|\sin u| + c$$

$$= \frac{1}{6} \ln\left|\sin(3x^2)\right| + c \ ,$$

(substituting u with $3x^2$).

19.   (C)

$(3x - 2)^3$ is a cubic polynomial, i.e., it is of the form
   $ax^3 + bx^2 + cx + d$.

Therefore, after three times differentiating, we will get a non-zero constant. But the derivative of a constant function is zero, so the fourth and higher derivatives will be zero.

Note: To see the answer, you only need to notice the degree. You do not have to do the actual computation.

20.   (E)

$$\frac{d}{dx}\left(\arctan(x)\right) = \frac{d}{dy}\left(\ln\left(y^2\right)\right) \cdot \frac{dy}{dx}$$

$$\frac{1}{1 + x^2} = \frac{2y}{y^2} \cdot \frac{dy}{dx}$$

$$\frac{y}{2(1 + x^2)} = \frac{dy}{dx}$$

21.   (E)

$$\frac{dv}{dt} = -40 + 2(.2)t$$

$$= -40 + 0.4t$$

Therefore, $\frac{dv}{dt}(40) = -40 + (0.4)(40)$

$$= -24 .$$

You should not change it to positive. It is negative, because V is a decreasing function.

22.   (C)

Average value of f in $[a,b]$ = $\frac{1}{b-a} \int_a^b f(x) \, dx$

$\Leftrightarrow 4 = \frac{1}{b-2} \cdot 8$, substituting the given values.

$\Leftrightarrow 4(b-2) = 8$

$\Leftrightarrow b = 4 .$

23.   (B)

$$\frac{d}{dx} \int_x^0 \frac{dt}{2+t}$$

$$= \frac{d}{dx}\left(-\int_0^x \frac{dt}{2+t}\right)$$

$$= -\frac{d}{dx}\left(\int_0^x \frac{dt}{2+t}\right), \text{ since } \frac{d}{dx}(cg)$$

$$= c \cdot \frac{dg}{dx} \text{ for } c \text{ a constant.}$$

$$= - \frac{1}{2 + x} \text{, since } \frac{1}{2 + t} \text{ is continuous in } [0, 1] \text{ and}$$

$$\frac{d}{dx} \int_{a}^{x} f(t) \, dt = f(x)$$

whenever f is continuous in $[a, b]$.

24.   (D)

$$\frac{x + x^2}{x^2 - 1} = \frac{x(1 + x)}{(x - 1)(x + 1)}$$

$$= \frac{x}{x - 1} \text{, for } x \neq 1.$$

Therefore, $\displaystyle\lim_{x \to -1} \frac{x + x^2}{x^2 - 1} = \lim_{x \to -1} \frac{x}{x - 1}$

$$= \frac{-1}{-1 - 1}$$

$$= \frac{1}{2}$$

<u>Remark</u>: The limit at $x = -1$ exists though it is not defined there.

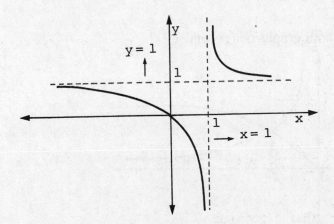

81

25. (C)

$$\frac{1}{h}\left(\frac{1}{x+h}-\frac{1}{x}\right)$$

$$=\frac{1}{h}\frac{(x-(x+h))}{(x+h)\,x}\,,\ \text{for } x\neq 0\,.$$

$$=\frac{1}{h}\frac{(-h)}{(x+h)x}$$

$$=-\frac{1}{(x+h)\,x} \qquad\qquad \text{.....................(i)}$$

Thus $\displaystyle\lim_{h\to 0}\ \frac{1}{h}\left(\frac{1}{x+h}-\frac{1}{x}\right)$

$$=\lim_{h\to 0}\ \left(\frac{-1}{x+h}\right)\left(\frac{1}{x}\right),\ \text{from (i) above}.$$

$$=-\frac{1}{x}\frac{1}{x}$$

$$=-\frac{1}{x^2}$$

Or, simply, if you observe, the limit desired is

$$\frac{d}{dx}\left(\frac{1}{x}\right)=-\frac{1}{x^2}$$

26. (D)

Draw both graphs of $f(x)$ and $f'(x)$.

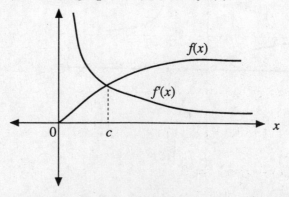

By tracing on the graph to the point where $x=c$, you can find that $c=0.49$.

27.   (B)

$$x(4) - x(0) = \int_0^4 V(t)\, dt$$

$$= \int_0^4 t^2\, dt + \int_2^4 (t+2)\, dt$$

$$= \frac{t^3}{3}\Big|_{t=0}^{t=2} + \left(\frac{t^2}{2} + 2t\right)\Big|_{t=2}^{t=4}$$

$$= \left(\frac{8}{3} - 0\right) + \left(\frac{16}{2} + 8\right) - \left(\frac{4}{2} + 4\right)$$

$$= 12\,\frac{2}{3}$$

Hence, $x(4) = 12\frac{2}{3} + x(0)$

$$= 12\frac{2}{3} + 0 \text{ , since it started at the origin.}$$

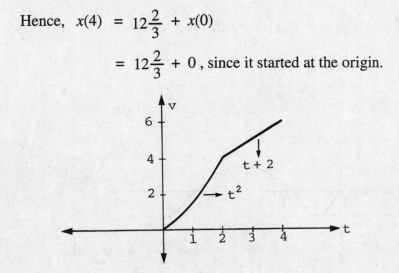

28.   (B)

$$f'(x) = g'(x)$$

$$\Leftrightarrow f'(x) - g'(x) = 0$$

$$\Leftrightarrow \frac{d}{dx}(f(x) - g(x)) = 0$$

$\Leftrightarrow$ $f(x) - g(x)$ is a constant, since the only function whose derivative is zero on an interval is a constant function.

But, $f(0) - g(0) = 2$. Thus, $f(x) - g(x) = 2$ throughout $[-1, 1]$.

Hence, $\displaystyle\int_{-1}^{1} [f(x) - g(x)] \, dx = \int_{-1}^{1} 2 \, dx$

$$= 2x \Big|_{-1}^{1}$$

$$= 2 - (-2)$$

$$= 4$$

29.   (A)

Draw the graph of $f(x) = \sin x \ln x$.

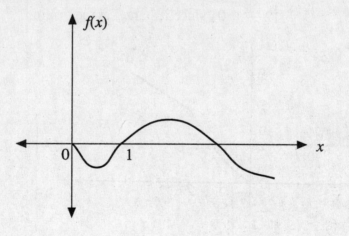

The area under the first arch is

$$\int_{0}^{1} \ln x \, \sin x.$$

So, *fnInt* $(\sin x \ln x, x, 0, 1)$ gives -0.2398.

30. (E)

$$\lim_{x \to \infty} \frac{(1 - 2x^2)^3}{(x^2 + 1)^3}$$

$$= \left(\lim_{x \to \infty} \frac{(1 - 2x^2)}{(x^2 + 1)} \right)^3$$

$$= (-2)^3$$

$$= -8$$

31. (D)

$$f'(x) = 1 - \frac{1}{x^2} \qquad \dots\dots\dots\dots\dots\dots(i)$$

$$\Leftrightarrow f'(x) = \frac{x^2 - 1}{x^2}$$

Therefore, $f'(x) = 0 \Leftrightarrow x = \pm 1$.

$f''(x) = \dfrac{d}{dx}(1 - x^{-2})$ differentiating (i) above.

$$= \frac{2}{x^3}$$

$\Rightarrow f''(1) = 2 > 0$ and $f''(-1) = -2 < 0$.

Hence, f has relative min. at $x = 1$.

32.   (B)

$$8 + 2x - x^2 = 0$$

$$\Leftrightarrow -(x^2 - 2x - 8) = 0$$

$$\Leftrightarrow x = 4 \quad \text{or} \quad x = -2$$

Since the coefficient of $x^2 < 0$, the graph of $y = 8 + 2x - x^2$ looks like the following:

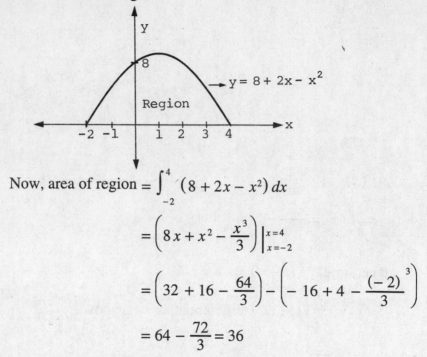

Now, area of region $= \displaystyle\int_{-2}^{4} (8 + 2x - x^2)\, dx$

$$= \left(8x + x^2 - \frac{x^3}{3} \right) \bigg|_{x=-2}^{x=4}$$

$$= \left(32 + 16 - \frac{64}{3} \right) - \left(-16 + 4 - \frac{(-2)^3}{3} \right)$$

$$= 64 - \frac{72}{3} = 36$$

Remark: You have to be careful in handling $\dfrac{-(-2)^3}{3}$ in the second set of parentheses: $(-2)^3 = -8$

$$\Rightarrow \frac{-(-2)^3}{3} = \frac{-(-8)}{3}$$

$$= \frac{8}{3}$$

If you make the error $\dfrac{-(-2)^3}{3} = -\dfrac{8}{3}$, you will get:

$$32 + 16 - \frac{64}{3} + 16 - 4 + \frac{8}{3} = 41\frac{1}{3}, \quad \text{which is incorrect.}$$

33. (A)

$$-1 \le x \le 1 \implies 0 \le x^2 \le 1$$

$$\implies x^2 - 1 \le 0$$

$$\implies |x^2 - 1| = -(x^2 - 1)$$

$$= 1 - x^2$$

Thus $\int_{-1}^{1} |(x^2 - 1)| \, dx = \int_{-1}^{1} (1 - x^2) \, dx$

$$= \left(x - \frac{x^3}{3} \right) \Big|_{-1}^{1}$$

$$= \frac{4}{3}$$

You have to take the same precaution as in problem 32.

34. (C)

Draw the graph of $f(x)$.

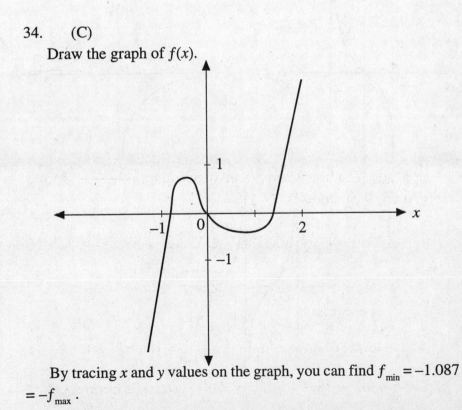

By tracing x and y values on the graph, you can find $f_{min} = -1.087$
$= -f_{max}$.

35.　(E)

A function is symmetric with respect to the origin if and only if $f(-x) = -f(x)$ for every real number x. You can also simply answer the question by looking at the graphs. As you will see below, for the first four graphs the reflection through the origin of a point $(x, f(x))$ is the point $(-x, f(-x))$:

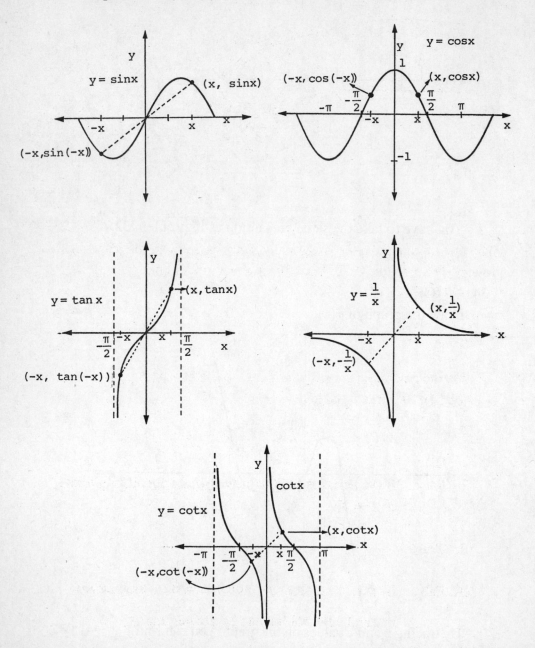

36. (A)

Draw the graphs of both y and y'.

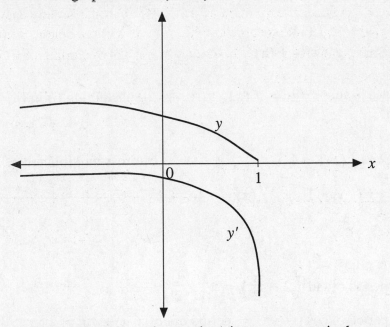

The rate of change, from the graph y', increases negatively as x goes toward 1. So, inside $0 \le x \le 0.8$, the highest rate of change is

$y'(0.8) = -1.14$.

37. (B)

Use the calculator directly to find $f'(2)$. For example, der 1 ($x^\wedge$ 0.5 ln x, x, 2) gives 0.95.

38. (B)

I. False: $f(x) = |x|$ is continuous at $x = 0$, but is not differentiable at $x = 0$.

II. True.

III. False: If $f(x) = 2$, then $\int_3^7 f(x)\,dx = 8$, which means integral does not have to be greater than 8.

39.   (D)

By definition of the derivative of an inverse function.

$$g'(x) = \frac{1}{f'(x)} \qquad\qquad\qquad \text{............(i)}$$

Since $g'(x) = f'(x)$, $\qquad\qquad\qquad$ ....(ii)

we have $\dfrac{1}{f'(x)} = f'(x),$ $\qquad\qquad$ substituting (i) in (ii)

$$\Leftrightarrow \left[f'(x)\right]^2 = 1, \qquad\qquad \text{by cross multiplication.}$$

$$\Leftrightarrow f'(x) = 1 \text{ or } f'(x) = -1.$$

40.   (C)

period of $\tan\left(2 - \dfrac{x}{2}\right) = \pi \div \dfrac{1}{2} = 2\pi$.

period of $\cos(x - \pi)$ = period of $\cos(-x + \pi) = 2\pi$.

$\dfrac{d}{dx}(\cos(-x + \pi)) = (-1)(-\sin(-x + \pi))$, by the chain rule.

$= \sin(-x + \pi)$

$$\Rightarrow \frac{d^2}{dx^2}(\cos(-x + \pi)) = \frac{d}{dx}\sin(-x + \pi)$$

$$= -\cos(-x + \pi), \text{ by the chain rule.}$$

Therefore, $\dfrac{d^2y}{dx^2} + y = \cos(-x + \pi) + (-1)\cos(-x + \pi)$

$$= 0$$

Moreover, if $f(x) = \cos(-x + \pi)$, then

$$f''(x) = -\cos(-x + \pi).$$

Therefore, $f''\left(-\dfrac{\pi}{2}\right) = -\cos\left(-\left(-\dfrac{\pi}{2}\right) + \pi\right)$

$$= -\cos\frac{3\pi}{2}$$

$$= 0$$

Also, $f''\left(\dfrac{\pi}{2}\right) = -\cos\left(-\dfrac{\pi}{2} + \pi\right)$

$$= -\cos\dfrac{\pi}{2}$$

$$= 0$$

Hence, it will have more than one inflection point.

ADVANCED PLACEMENT CALCULUS AB EXAM II

SECTION II

DETAILED EXPLANATIONS OF ANSWERS

1.

$0 \leq f(x) \leq 1$, given. ....(i)

$g(0) = 0 - f(0) \leq 0$, since $f(0) \geq 0$.

$\Leftrightarrow g(0) = 0$ or $g(0) < 0$ . ...(ii)

$g(1) = 1 - f(1) \geq 0$, since $f(1) \leq 1$. ....(iii)

$\Leftrightarrow g(1) = 0$ or $g(1) > 0$. ....(iv)

If $g(0) = 0$ or $g(1) = 0$, then

$0 - f(0) = 0$ or $1 - f(1) = 0$

$\Leftrightarrow f(0) = 0$ or $f(1) = 1$ ....(v)

If (v) holds, then $c = 0$ or $c = 1$ will give $f(c) = c$.

But if neither $g(0) = 0$ nor $g(1) = 0$, then from (ii) and (iv) we

get $g(0) < 0$ and $g(1) > 0$. Hence, by the intermediate value theorem, there is a '*c*' such that

$$g(c) = 0$$

$$\Leftrightarrow c - f(c) = 0$$

$$\Leftrightarrow f(c) = c .$$

Alternatively, the graph of f should take one of the following forms:

If f does not start at the origin, then its graph will be like those in (A), (C) or (F). If f does not touch the point (1,1), then its graph will resemble (B), (C), or (D); otherwise it will be like (E). In either case the graph of $y = f(x)$ will intersect the graph of $y = x$ at some $x = c$.

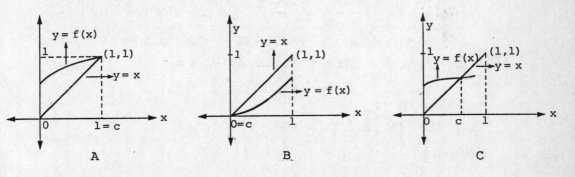

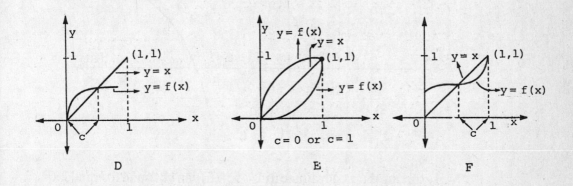

2.   (A)
Draw the graph of $y = x^2 - x - 6$.

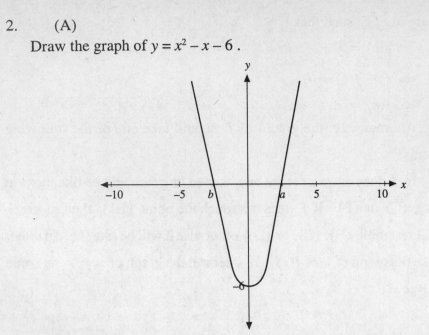

The function $(\ln x^{\wedge 2} - x - 6)$ requires $x^2 - x - 6 > 0$. By tracing down the graph, you can easily see that $b = -2$, $a = 3$.

(B)
Draw the graph $f(x) = \ln(x^2 - x - 6)$ for $x \geq 3$.

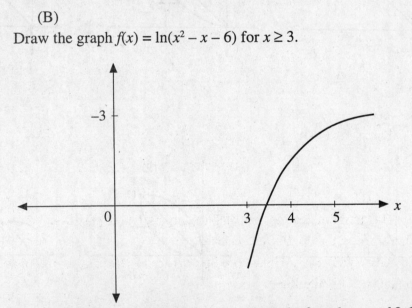

By tracing the x coordinate to $x = 5$, $f(5)$ can be found to equal 2.64.

(C)
Use your calculator directly and perform
der 1 $(\ln x^{\wedge} 2 - x - 6, x, -3)$, which should give -1.67.

94

3.   (A)

Domain $= R$, since $x^2 + 1 \neq 0$, for all real x.

(B)

$$\frac{1 - x^2}{x^2 + 1} = \frac{1 - x^2}{x^2} \div \frac{x^2 + 1}{x^2},$$ dividing both numerator and denominator by x^2

$$= \left(\frac{1}{x^2} - 1\right) \div \left(1 + \frac{1}{x^2}\right)$$

Therefore, $\lim\limits_{x \to \pm\infty} \left(\frac{1 - x^2}{x^2 + 1}\right) = \lim\limits_{x \to \pm\infty} \left[\left(\frac{1}{x^2} - 1\right) \div \left(1 + \frac{1}{x^2}\right)\right]$

$$= -1 \div 1$$

$$= -1$$

(C)

$$f'(x) = -\frac{4x}{(x^2 + 1)^2}, \quad \text{by the quotient rule.} \quad \dots\dots\dots\dots(i)$$

Also, $(x^2 + 1)^2 > 0$, for any x,

$-4x > 0$, when $x < 0$,

$-4x < 0$, when $x > 0$.

Hence, $\quad -\dfrac{4x}{\left(x^2 + 1\right)^2} \quad > 0$, when $x < 0$,

and $\quad -\dfrac{4x}{\left(x^2 + 1\right)^2} \quad < 0$, when $x > 0$.

Therefore, $f'(x) > 0$ when $x < 0$, and

$$f'(x) < 0 \text{ when } x > 0.$$

(D)

The tangent line is parallel to the x-axis so its slope is zero. Therefore, $f'(x) = 0 \Leftrightarrow x = 0$.

Also, $f(0) = 1$. Hence the line parallel to the x-axis and tangent to f at $(0, 1)$ is the line $y = 1$.

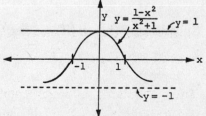

4.   **(A)**

Let $h(t)$ = Height at time t ,

$r(t)$ = Radius at time t .

Therefore, the volume at time t , $V(t) = \frac{1}{3}\,\pi r(t)^2 h(t)$.

$\dfrac{dV}{dt} = 2$ in³/ min, given.   ....(i)

But, from similarity of triangles $\dfrac{h(t)}{r(t)} = \dfrac{12}{4} = 3$

$\Leftrightarrow h(t) = 3r(t)$

$\Leftrightarrow r(t) = \frac{1}{3}h(t)$

Thus $V(t) = \frac{1}{3}\,\pi\left(\frac{1}{3}h\right)^2 h(t)$

$= \dfrac{1}{27}\,\pi h^3$ ....(ii)

$\Rightarrow \dfrac{dV}{dt} = \dfrac{1}{9}\,\pi h^2\,\dfrac{dh}{dt}$

$\Rightarrow 2 = \dfrac{1}{9}\,\pi h^2\,\dfrac{dh}{dt}$, by (i) and (ii).

$\Rightarrow \dfrac{dh}{dt} = \dfrac{18}{h^2\pi}$.   ...(iii)

Therefore, when $h = 6$, $\dfrac{dh}{dt} = \dfrac{18}{6^2\pi}$ inches/min.

$= \dfrac{1}{2\pi}$ inches/min.

(B)

When the cup is full its volume $= \frac{1}{3}\,\pi 4^2 \cdot 12$

$= 64\,\pi$ in³

Therefore, when it is half full $V(t) = 32\pi$ in³ .   ...(iv)

Substituting (iv) in (ii) we get: $32\pi = \dfrac{1}{27}\,\pi h^3$

$\Leftrightarrow h = 6\sqrt[3]{4}$.   ...(v)

Finally using (i), (iii) and (v):

$$\frac{dh}{dt} = \frac{18}{\pi \left(6\sqrt[3]{4}\right)^2} \quad \text{inches/min.}$$

$$= \frac{1}{2\pi \sqrt[3]{4^2}} \quad \text{inches/min.}$$

5. (A)
$$\frac{d^2}{dx^2}(-2f) = -2f''.$$

From the chart, $-2f'' > 0$ in $(-\infty, 0)$,
$$-2f'' < 0 \text{ in } (0,\infty).$$
Therefore, $-2f$ has inflection point at $x = 0$.

(B) (i)
$$\frac{d}{dx}(-2f) = -2f' < 0 \text{ in } (-\infty, -2) \text{ from the chart.}$$
$$-2f' > 0 \text{ in } (-2, 0) \text{ from the chart.}$$
$\Rightarrow -2f$ has relative min. at $x = -2$.

(ii)
By similar argument as in (i), $-2f$ has relative max. at $x = 2$ and equals 0.

(C)
Again from the table, $-2f'' > 0$ in $(-\infty, 0)$, and
$$-2f'' < 0 \text{ in } (0,\infty).$$
So, it is concave upward in $(-\infty,0)$ and concave downward in $(0,\infty)$

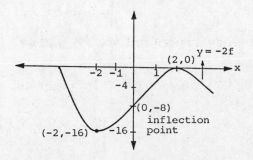

6.　(A)

$y = x$ and $y = x^3$ intersect

$\Leftrightarrow x = x^3$

$\Leftrightarrow x - x^3 = 0$

$\Leftrightarrow x\,(1 - x^2) = 0$

$\Leftrightarrow x = -1, x = 0, x = 1.$

(B)

Determine which lies above the other:

x	x		x^3
$-\dfrac{1}{2}$	$-\dfrac{1}{2}$	$<$	$-\dfrac{1}{8}$
$\dfrac{1}{2}$	$\dfrac{1}{2}$	$>$	$\dfrac{1}{8}$

Therefore, $y = x^3$ lies above $y = x$ in $(-1,0)$ and $y = x$ lies above $y = x^3$ in $(0,1)$.

$$\text{Area} = \int_{-1}^{0} (x^3 - x)\, dx + \int_{0}^{1} (x - x^3)\, dx$$

$$= \left(\frac{x^4}{4} - \frac{x^2}{2}\right)\Big|_{-1}^{0} + \left(\frac{x^2}{2} - \frac{x^4}{4}\right)\Big|_{0}^{1}$$

(C)

$$\text{Volume} = \int_{-1}^{0} \pi\,(x^2 - x^6)\, dx + \pi\int_{0}^{1} (x^2 - x^6)\, dx,$$

since $x^2 > x^6$ in $(-1,0)$; look at the figure.

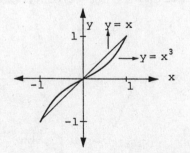

98

Remark: Do not write $\int_{-1}^{0} (x - x^3)\, dx$, since area is non-negative.

Also, do not write $\pi \int_{-1}^{0} (x^6 - x^2)\, dx$, since volume is non-negative.

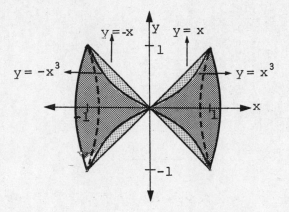

THE ADVANCED PLACEMENT EXAMINATION IN

CALCULUS AB

TEST III

ADVANCED PLACEMENT CALCULUS AB EXAM III

SECTION I

PART A

Time: 45 minutes
25 questions

DIRECTIONS: Each of the following problems is followed by five choices. Solve each problem, select the best choice, and blacken the correct space on your answer sheet. Calculators may not be used for this section of the exam.

NOTE:
Unless otherwise specified, the domain of function f is assumed to be the set of all real numbers x for which $f(x)$ is a real number.

1.   For what value of x will the tangent lines to $y = \ln x$ and $y = 2x^2$ be parallel?

(A) 0   (D) 1

(B) $\dfrac{1}{4}$ (E) 2

(C) $\dfrac{1}{2}$

2.   If $f(x) = 2^{x^3+1}$, then $f'(1)$ is approximately:

(A) 2.000   (D) 2.773

(B) 4.000   (E) 8.318

(C) 6.000

3. Let $f'(x) = \sin(\pi x)$ and $f(0) = 0$. Then $f(1) = ?$

(A) $-\dfrac{1}{\pi}$

(B) $\dfrac{1}{\pi}$

(C) $-\dfrac{2}{\pi}$

(D) $\dfrac{2}{\pi}$

(E) None of these

4. Let the velocity of a point moving on a line at a time t be defined by $V(t) = 2^t \ln 2$ cm/sec. How many centimeters did the point travel from $t = 0$ sec. to $t = 2.5$ sec.?

(A) 8.882

(B) 9.882

(C) 4.104

(D) 8.121

(E) 10.003

5. Find the slope of the tangent line to the graph of

$$y = \frac{5}{4+x^3} \text{ when } x = 1.$$

(A) 8.047

(B) −0.600

(C) −0.200

(D) 1.000

(E) None of the above

6. Let $f(x) = e^{bx}$, $g(x) = e^{ax}$ and find the value of b such that

$$D_x \left(\frac{f(x)}{g(x)} \right) = \frac{f'(x)}{g'(x)}$$

(A) $\dfrac{a^2}{a^2 - 1}$

(D) $\dfrac{a - 1}{a^2}$

(B) $\dfrac{a^2}{a + 1}$

(E) $\dfrac{a^2}{a - 1}$

(C) $\dfrac{a + 1}{a^2}$

7. Suppose $x^2 - xy + y^2 = 3$. Find $\dfrac{dy}{dx}$ at the point (a,b).

(A) $\dfrac{a - 2b}{2a - b}$

(D) $\dfrac{b - 2a}{2b + a}$

(B) $\dfrac{b - 2a}{2b - a}$

(E) $\dfrac{b + 2a}{2b + a}$

(C) $\dfrac{a - 2b}{2a + b}$

8. $\lim\limits_{h \to 0} \dfrac{e^{x+h} - e^x}{h}$ equals

(A) 0

(D) $-\infty$

(B) 1

(E) None of these

(C) $+\infty$

105

9. $\int_{-2}^{-1} x^{-4}\,dx = ?$

(A) $\dfrac{7}{2}$

(D) $\dfrac{31}{160}$

(B) $\dfrac{31}{8}$

(E) None of these

(C) $\dfrac{7}{24}$

10. If $f'(x) = \dfrac{x^2}{2}$ where $f(0) = 0$ then $3f(4) =$

(A) 0

(D) 24

(B) 3

(E) 32

(C) 12

11. $\displaystyle \lim_{x \to +\infty} \left(\dfrac{1}{x} - \dfrac{x}{x-1} \right) = ?$

(A) -1

(D) 2

(B) 0

(E) None of these

(C) 1

12. Let $R = \int_0^a \cos(x^2)\, dx$ and $S = \int_0^a \tan x\, dx$

Find $\int_{-a}^a [\cos(x^2) + \tan x]\, dx$.

(A) $2R$ (D) $S + 2R$

(B) $2S$ (E) $2R + 2S$

(C) $R + 2S$

13. If $\sin y = \cos x$, then find $\dfrac{dy}{dx}$ at the point $(\frac{\pi}{2}, \pi)$.

(A) -1 (D) $\dfrac{\pi}{2}$

(B) 0 (E) None of these

(C) 1

14. For which of the following intervals is the graph of $y = x^4 - 2x^3 - 12x^2$ concave down?

(A) $(-2, 1)$ (D) $(-\infty, -1)$

(B) $(-1, 2)$ (E) $(-1, +\infty)$

(C) $(-1, -2)$

15. $\int_1^e x \ln x \, dx = ?$

(A) e

(D) $\dfrac{e-1}{2}$

(B) $\dfrac{e^2-1}{2}$

(E) None of these

(C) $\dfrac{e^2+1}{4}$

16. If $f'(x) = 2(3x+5)^4$, then the fifth derivative of $f(x)$ at $x = -\dfrac{5}{3}$ is

(A) 0

(D) 3888

(B) 144

(E) None of these

(C) 1296

17. Let $f(x) = \dfrac{x}{\sqrt{x^2-8}}$. Which of the following interval notations represents the most inclusive domain for f?

(A) $(2\sqrt{2}, -2\sqrt{2})$

(B) $[2\sqrt{2}, -2\sqrt{2}]$

(C) $(-\infty, +\infty)$

(D) $(-\infty, -2\sqrt{2}] \cup [2\sqrt{2}, +\infty)$

(E) $(-\infty, -2\sqrt{2}) \cup (2\sqrt{2}, +\infty)$

18. If $f(x) = \ln x$ then $f\left(\dfrac{3}{2}\right) = ?$

(A) $\dfrac{\ln 3}{\ln 2}$

(D) $\displaystyle\int_2^3 \ln t \; dt$

(B) $\ln 2 - \ln \dfrac{1}{2}$

(E) $\displaystyle\int_2^3 \dfrac{1}{t} \; dt$

(C) $\displaystyle\int_{\ln 2}^{\ln 3} e^t \; dt$

19. If $y = \dfrac{3}{\sin x + \cos x}$ then $\dfrac{dy}{dx} =$

(A) $3 \sin x - 3 \cos x$

(B) $\dfrac{6 \sin x}{1 + 2 \sin x \; \cos x}$

(C) $\dfrac{3}{\cos x - \sin x}$

(D) $\dfrac{-3}{(\sin x + \cos x)^2}$

(E) $\dfrac{3(\sin x - \cos x)}{1 + 2 \sin x \; \cos x}$

20. $\displaystyle\int_{-2}^{-1} \left| x^{-3} \right| dx =$

(A) $\dfrac{3}{8}$

(D) $\dfrac{15}{64}$

(B) $\dfrac{5}{8}$

(E) None of these

(C) $\dfrac{15}{4}$

109

21. $\displaystyle\lim_{x \to 0} \frac{\dfrac{3}{x^2}}{\dfrac{2}{x^2} + \dfrac{105}{x}} =$

(A) 0

(D) $\dfrac{3}{107}$

(B) 1

(E) None of these

(C) $\dfrac{3}{2}$

22. The graph of f is shown in the figure. Which of the following could be the graph of $\int f(x)\, dx$?

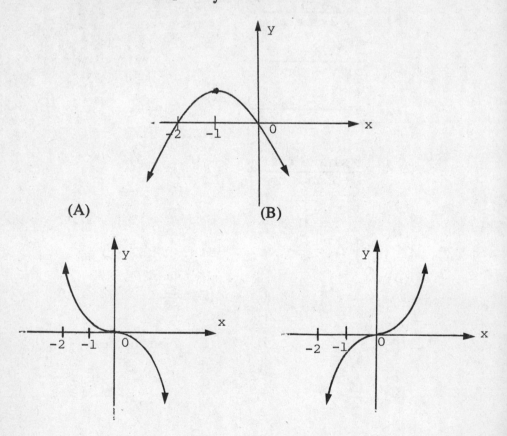

(A)

(B)

(C)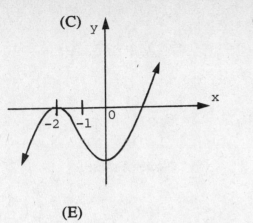

(D)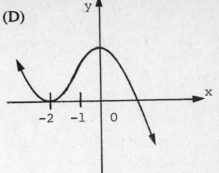

(E)

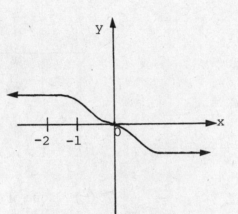

23. If $f(x) = \int (1 - 2x)^3 \, dx$, then the second derivative of

$f(x)$ at $x = \dfrac{1}{2}$ is

(A) -48 (D) 96

(B) -12 (E) None of these

(C) 0

24. Let $F(x) = \int_1^x f(t)\, dt$, and use the graph given of $f(t)$ to find $F'(1) =$

(A) 0

(D) $\dfrac{1}{2}$

(B) 1

(E) None of these

(C) 2

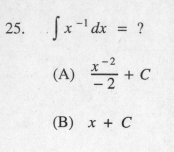

y = f(t)

25. $\int x^{-1} dx = ?$

(A) $\dfrac{x^{-2}}{-2} + C$

(D) $-x + c$

(B) $x + C$

(E) None of these

(C) Undefined

PART B

Time: 45 minutes
15 questions

DIRECTIONS: Calculators may be used for this section of the test. Each of the following problems is followed by five choices. Solve each problem, select the best choice, and blacken the correct space on your answer sheet.

NOTES:

1. Unless otherwise specified, answers can be given in unsimplified form.

2. The domain of function f is assumed to be the set of all real numbers x for which $f(x)$ is a real number.

26. The area that is enclosed by $y = x^3 + x^2$ and $y = 6x$ for $x \geq 0$ is

(A) $\dfrac{29}{12}$ (D) 6

(B) 3   (E) $\dfrac{32}{3}$

(C) $\dfrac{16}{3}$

27. $\lim\limits_{x \to 0} \dfrac{\arctan x}{\tan x}$ equals

(A) 0

(B) 1

(C) $+\infty$

(D) $-\infty$

(E) None of these

28. Find the area in the first quadrant that is enclosed by $y = \sin 3x$ and the x-axis from $x = 0$ to the first x-intercept on the positive x-axis.

(A) $\dfrac{1}{3}$

(B) $\dfrac{2}{3}$

(C) 1

(D) 2

(E) 6

29. A particle moves along a straight line. Its velocity is

$$V(t) = \begin{cases} t^2 & \text{for } 0 \le t \le 2 \\ t + 2 & \text{for } t \ge 2 \end{cases}$$

The distance travelled by the particle in the interval $1 \le t \le 3$ is

(A) 3

(B) 7.1

(C) 4.3

(D) 5

(E) 6.8

30. Let $f(x) = x + \dfrac{1}{x^{1.6}}$. Then, $\displaystyle\int_{0.1}^{2} f(x)$ equals

    (A) 3.64   (D) 5.1

    (B) 4.99   (E) 11.2

    (C) 7.53

31. $\displaystyle\lim_{x \to -4^{+}} \dfrac{4x - 6}{2x^2 + 5x - 12}$ equals

    (A) 0   (D) $-\infty$

    (B) 1   (E) None of these

    (C) $+\infty$

32. Let $g(x) = e^{-x^2}$ and determine which one of the following statements is true.

    (A) g is a decreasing function.

    (B) g is an odd function.

    (C) g is symmetric with respect to the x-axis.

    (D) $(.5, e^{-.25})$ is a point of inflection.

    (E) None of these.

33. The graph shown represents $y = f(x)$. Which one of the following is NOT true?

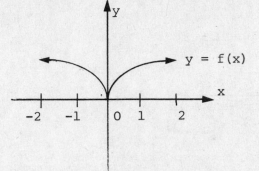

(A) f is continuous on $(-2, 2)$.

(B) $\lim\limits_{x \to 0} f(x) = f(0)$.

(C) f is differentiable on $(-2, 2)$.

(D) $\lim\limits_{x \to 0^-} f(x) = f(0)$.

(E) $f'(x) < 0$ for $x < 0$.

34. Which one of the following is NOT an antiderivative of $\sec x$?

(A) $\ln|\sec x + \tan x| + C$

(B) $-\ln|\sec x - \tan x| + C$

(C) $\ln\left|\dfrac{1 - \sin x}{\cos x}\right| + C$

(D) $\ln\left|\dfrac{1 + \sin x}{\cos x}\right|$

(E) $\ln\left|\dfrac{\cos x}{1 - \sin x}\right|$

35. If $8 + 2x - x^2 = 6x^3$, one solution for x is

(A) 1.143 (D) 5.78

(B) 2.25 (E) 12.3

(C) −17.2

36. Let $f(x) = \dfrac{\sin x \cos x}{\cos 9x \tan 2x}$. $f'(0.5)$ equals

(A) 3.8 (D) 70.1

(B) −2.5 (E) −5

(C) −49.5

117

37. Let $f(x) = \dfrac{x}{\sqrt{4 - x^2}}$. The minimum of $f'(x)$ is

(A) 0.5

(D) −0.5

(B) 1

(E) 2

(C) −1

38. Suppose $h(x)$ is a continuous function on $[1, 2]$ and $f(1) = 2$, $f(1.5) = 0.5$, and $f(2) = -3$. Which of the following is FALSE?

(A) The maximum value of f in $[1, 2]$ is 2.

(B) $f(c) = 0$ for some real value of c.

(C) $\lim\limits_{x \to 2^-} f(x) = -3$

(D) $\left|f(2)\right| - \left|f(1)\right| \le \left|f(2) - f(1)\right|$

(E) $\lim\limits_{x \to 5/4} f(x) = f\left(\dfrac{5}{4}\right)$

39. Find the value of c such that the area between the line $y = c$ and the parabola $y = x^2$ is $\dfrac{1}{48}$.

(A) $\dfrac{1}{512}$

(D) $\dfrac{1}{16}$

(B) $\dfrac{1}{64}$

(E) None of these

(C) $\dfrac{1}{32}$

40. Let f', g' be differentiable functions that are derivatives of f and g, respectively. If $f'(x) \le g'(x)$ for all real x, which of the following must be true?

I. $\displaystyle\lim_{x \to a} f'(x) \le \lim_{x \to a} g'(x)$

II. $\displaystyle\int_a^b f'(x)\,dx \le \int_a^b g'(x)\,dx$

III. $f(x) \le g(x)$, for all real x

IV. $\displaystyle\int_a^b f(x)\,dx \le \int_a^b g(x)\,dx$

(A) I only

(B) II only

(C) I and II only

(D) I, II, and III

(E) I, II, III, and IV

119

ADVANCED PLACEMENT
CALCULUS AB EXAM III

SECTION II

Time: 1 hour and 30 minutes
        6 problems

DIRECTIONS: Show all your work. Grading is based on the methods used to solve the problem as well as the accuracy of your final answers. Please make sure all procedures are clearly shown.

NOTES:

1.   Unless otherwise specified, answers can be given in unspecified form.

2.   The domain of function f is assumed to be the set of all real numbers x for which $f(x)$ is a real number.

1.   Let f be the function defined by $f(x) = \dfrac{x}{(x-1)^2}$

   (A)   Sketch the graph of $y = f(x)$. Be sure to label all relative extrema, points of inflection, and asymptotes.

   (B)   Use interval notation to indicate where the function is increasing, decreasing, concave up, and concave down.

   (C)   State the domain and range of the function and indicate why the function does not have an inverse over its domain.

   (D)   Write the equation (in standard form) for the tangent line to $y = f(x)$ that passes through a point of inflection.

2.

   Let $f(x) = \dfrac{\dfrac{3}{x^2}}{\dfrac{2}{x^2} + \dfrac{105}{x}}$

(A) Find $f'(0)$.

(B) What happens at $x = -0.019$? Explain.

(C) Find $\int_0^2 f(x)$.

3.   Let the growth of a plant culture X (measured in grams/sec) be directly proportional to the weight of the plant culture (measured in grams).

(A) Write a formula for the growth of the plant culture as an instantaneous rate of change of weight in terms of the weight of the plant culture.

(B) Use the formula in part (A) to write a formula for the weight of the plant culture X as a function of time.

(C) If the weight of the culture at the beginning of the experiment ($t = 0$) is 10 grams, and the weight after one second ($t = 1$) is 100 grams, find the weight of the culture after 0.5 seconds.

(D) Find the difference between the instantaneous rate of growth when t is one second and the average rate of growth for the first second.

4. Let $y = \cos(2x^2)$ for x in the interval $[0, c]$ where c is the first positive x - intercept.

    (A) Find the volume of the region between $y = f(x)$ and $y = 0$ in the interval $[0, c]$, rotating about the y -axis.

    (B) Find the average value of $y = f'(x)$ over the given interval.

    (C) Write (DO NOT EVALUATE) the definite integral expression for the volume of the region between $y = 1$ and $y = f(x)$ on the given interval, rotating about the x -axis.

5. Assume the volume of a cube is increasing at a constant rate of 3 cm^3 per second. Let t_0 be the instant when the rate of change of the volume (cm^3/sec) is equal to the rate of change of the surface area (cm^2/sec) for the cube. Assume $V = 0$ when $t = 0$.

    (A) Find the value(s) of t_0.

    (B) Find the rate of change of a side when $t = t_0$.

    (C) Find the rate of change of the surface area when $t = t_0$.

6.   The velocity of a glider (ft/min) is indicated by the following graph.

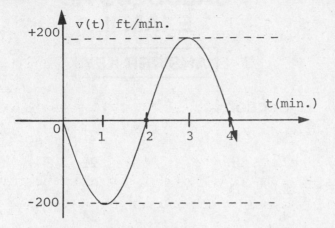

(A)   Let $t \in [0,4]$ and sketch a possible graph of $y = f(t)$, where y is the altitude in feet, t is the time in minutes and $f(4) = 1.5f(0) = 6000$ feet.

(B)   For what value(s) of t will the glider be at a minimum altitude for $t \in [0,4]$? Justify your conclusion.

(C)   For what value(s) of t will the glider be at a maximum altitude for $t \in [0,4]$? Justify your conclusion.

(D)   For what value(s) of t is the graph of $y = f(t)$ concave up? Justify your conclusion.

ADVANCED PLACEMENT CALCULUS AB EXAM III

ANSWER KEY

Section I

1.	C		21.	C
2.	E		22.	D
3.	D		23.	C
4.	A		24.	C
5.	B		25.	E
6.	E		26.	C
7.	B		27.	B
8.	E		28.	B
9.	C		29.	E
10.	E		30.	C
11.	A		31.	C
12.	A		32.	E
13.	C		33.	C
14.	B		34.	C
15.	C		35.	A
16.	D		36.	C
17.	D		37.	A
18.	E		38.	A
19.	E		39.	D
20.	A		40.	C

Section II

See Detailed Explanations of Answers.

ADVANCED PLACEMENT CALCULUS AB EXAM III

SECTION I

DETAILED EXPLANATIONS OF ANSWERS

1.   (C)

$y = \ln x$ and $y = 2x^2$

Set the derivatives equal and get

$y' = \dfrac{1}{x}$ and $y' = 4x$

$\dfrac{1}{x} = 4x$

$1 = 4x^2$

$\dfrac{1}{4} = x^2$ $\boxed{\text{so}\quad x = \pm\dfrac{1}{2}}$

Use $x = \dfrac{1}{2}$ since $y = \ln x$ is not defined for $x = -\dfrac{1}{2}$

2.   (E)

If $f(x) = 2^{x^3+7}$ then $f'(x) = 2^{x^3+1}(\ln 2)D_x\left(x^3+1\right)$

$= 2^{x^3+1}(\ln 2)\left(3x^2\right)$

So $f'(1) = 2^2(\ln 2)(3) = 12\ln 2 = \approx 8.318$

3.   (D)

$y = f(x) = \int \sin(\pi x)\, dx$

$= -\frac{1}{\pi}\cos(\pi x) + C$

Now substitute $x = 0$.

$0 = -\frac{1}{\pi}\cos 0 + C$, so $C = \frac{1}{\pi}$ and

$f(x) = -\frac{1}{\pi}\cos(\pi x) + \frac{1}{\pi}$.

$f(1) = -\frac{1}{\pi}\cos \pi + \frac{1}{\pi}$

$= -\frac{1}{\pi}(-1) + \frac{1}{\pi}$

$= \frac{2}{\pi}$

4.   (A)

$V(t) = 2^t \ln 2$, so $s(t) = 2^t + c$

$s(o) = 2^0 + c$ and

$$s(2.75) = 2.5^{2.5} + c$$
$$\approx 9.882 + c$$

$$\begin{aligned}
\text{distance} &= s(2.5) - s(0) \\
&= (9.882 + c) - (1 + c) \\
&= 8.882 \text{ cm.}
\end{aligned}$$

5.   (B)

$$y' = 5 \bullet (-1)\left(4 + x^3\right)^{-2}\left(3x^2\right)$$
$$= -5(5)^{-2}(3)$$
$$= -\frac{15}{25} = -\frac{3}{5}$$
$$= -0.600$$

6.   (E)

$$D_x\left(\frac{f(x)}{g(x)}\right) = \frac{e^{ax} b \, e^{bx} - e^{bx} a \, e^{ax}}{e^{2ax}}$$

$$= \frac{(b - a) \, e^{bx}}{e^{ax}}$$

$$\frac{f'(x)}{g'(x)} = \frac{b \, e^{bx}}{a \, e^{ax}}.$$

Now equate the two to get

$$\frac{b}{a} = b - a \quad \text{or} \quad b = ab - a^2 \Rightarrow$$

$$a^2 = ab - b \Rightarrow a^2 = b\,(a - 1)$$

Therefore $\dfrac{a^2}{a - 1} = b$

7. (B)

Differentiate $x^2 - xy + y^2 = 3$ implicitly to get

$2x - (xy' + y) + 2yy' = 0$.

Now solve for y' :

$2x - xy' - y + 2yy' = 0$

$y'(2y - x) = y - 2x$

$y' = \dfrac{y - 2x}{2y - x}$.

Let $x = a$, $y = b$ to get $y' = \dfrac{b - 2a}{2b - a}$

8. (E)

$\lim\limits_{h \to 0} \dfrac{e^{x+h} - e^x}{h}$ equals the derivative of e^x by using the definition of the derivative

$$f'(x) = \lim\limits_{h \to 0} \dfrac{f(x + h) - f(x)}{h}$$

$$= e^x \text{ for } f(x) = e^x$$

9. (C)

$$\int_{-2}^{-1} x^{-4} dx = \dfrac{x^{-3}}{-3}\Big|_{-2}^{-1}$$

$$= -\dfrac{1}{3}\left(\dfrac{1}{-1} - \dfrac{1}{-8}\right)$$

$$= -\dfrac{1}{3}\left(-\dfrac{7}{8}\right)$$

$$= \dfrac{7}{24}$$

10. (E)

$$f(x) = \int \frac{x^2}{2} \, dx$$

$$= \frac{x^3}{6} + C$$

Let $x = 0$; $f(0) = \frac{1}{6}(0)^3 + C \qquad 0 = C$

$$f(x) = \frac{1}{6} x^3$$

$$3 f(4) = 3 \cdot \frac{1}{6} \cdot (4)^3$$

$$= 32$$

11. (A)

$$\lim_{x \to +\infty} \left(\frac{1}{x} - \frac{x}{x-1} \right) = \lim_{x \to +\infty} \frac{x - 1 - x^2}{x \, (x-1)}$$

$$= \lim_{x \to \infty} \frac{-x^2 + x - 1}{x^2 - x}$$

$$= \lim_{x \to \infty} \frac{-1 + \frac{1}{x} - \frac{1}{x^2}}{1 - \frac{1}{x}}$$

$$= -1$$

12. (A)

$\cos(x^2)$ is an even function and $\tan x$ is an odd function so

$$\int_{-a}^{a} \cos(x^2) \, dx = 2R \ ; \quad \int_{-a}^{0} \tan x \, dx = -\int_{0}^{a} \tan x \, dx$$

Then $\int_{-a}^{a} [\cos(x^2) + \tan x] \, dx = 2R + 0$

$$= 2R$$

129

13. (C)

Differentiate $\sin y = \cos x$ implicitly.

$$\frac{dy}{dx} \cos y = -\sin x \ , \quad \frac{dy}{dx} = -\frac{\sin x}{\cos y} \ .$$

At $\left(\frac{\pi}{2}, \pi\right), \frac{dy}{dx} = -\frac{\sin\left(\frac{\pi}{2}\right)}{\cos(\pi)}$

$$= \frac{-1}{-1}$$

$$= 1$$

14. (B)

$$y = x^4 - 2x^3 - 12x^2$$

$$y' = 4x^3 - 6x^2 - 24x$$

$$y'' = 12x^2 - 12x - 24$$

$$= 12(x - 2)(x + 1)$$

```
+ + + + 0 - - - - - - - - - - 0 + + +
<----·--·--+--·--·--+--·--·--+--·--·--+--·--·-->
        -1   (concave down)   2
```

15. (C)

Use integration by parts with

$u = \ln x$	$du = \frac{1}{x} dx$
$dv = x$	$v = \frac{x^2}{2}$

$$\int_1^e x \ln x \ dx = \frac{x^2}{2} \ln x - \int_1^e \frac{x}{2} dx$$

$$= \left(\frac{x^2}{2} \ln x - \frac{x^2}{4}\right)\Big|_1^e$$

$$= \frac{e^2 + 1}{4}$$

16. **(D)**

If $f'(x) = 2(3x+5)^4$; $f''(x) = 24(3x+5)^3$

$f'''(x) = 216(3x+5)^2$; $f^{(4)}(x) = 1296(3x+5)$

$f^{(5)}(x) = 3888$ so $f\left(-\dfrac{5}{3}\right) = 3888$

17. **(D)**

If $f'(x) = \dfrac{x}{\sqrt{x^2-8}}$ then $f(x) = \int x(x^2-8)^{-1/2}\, dx$

Let $u = x^2 - 8$ and $du = 2x\, dx$, so

$f(x) = \dfrac{1}{2}\int u^{-1/2}\, du$

$\qquad = \dfrac{1}{2} \cdot 2u^{1/2} + C$

$\qquad = \sqrt{x^2 - 8} + C$.

We must have $|x^2 - 8| > 0$ in order for f to be real-valued so

$(x + 2\sqrt{2})(x - 2\sqrt{2}) > 0$

```
+ + + + +0 - - - - - -0 + + + + +
         ●            ●
      -2√ 2         2√2
```

$x \leq -2\sqrt{2}$ or $x \geq 2\sqrt{2}$,

which is written in interval notation as

$(-\infty, -2\sqrt{2}] \cup [2\sqrt{2}, +\infty)$

18.   (E)

Recall that $e^x = \int_1^x \frac{1}{t} dt$, so $\ln \frac{3}{2}$

$$= \ln 3 - \ln 2$$

$$= \int_1^3 \frac{1}{t} dt - \int_1^2 \frac{1}{t} dt$$

$$= \int_2^3 \frac{1}{t} dt$$

19.   (E)

$$y = \frac{3}{\sin x + \cos x} = 3u^{-1} \text{ where } u = \sin x + \cos x$$

So $\frac{dy}{dx} = 3(-1)u^{-1-1} \frac{du}{dx}$ using the derivative of a power and

chain rule theorems.

Thus $\dfrac{dy}{dx} = \dfrac{-3}{(\sin x + \cos x)^2}(\cos x - \sin x)$

$$= \frac{3(\sin x - \cos x)}{\sin^2 x + 2\sin x \cos x + \cos^2 x}$$

$$\frac{dy}{dx} = \frac{3(\sin x - \cos x)}{1 + 2\sin x \cos x} ,$$

using the identity $\sin^2 x + \cos^2 x = 1$

20.   (A)

Note $|n| = \begin{cases} n \text{ if } n \geq 0 \\ -n \text{ if } n < 0 \end{cases}$ so $|x^{-3}| = -x^{-3}$ for x in $[-2, -1]$

Now $\displaystyle \int_{-2}^{-1} \left| x^{-3} \right| dx = \int_{-2}^{-1} - x^{-3}\, dx$

$$= -\left(\frac{x^{-3+1}}{-3+1} \right)\Big|_{-2}^{-1}$$

$$= \frac{x^{-2}}{2}\Big|_{-2}^{-1}$$

$$= \frac{1}{2}\left[\frac{1}{(-1)^2} - \frac{1}{(-2)^2} \right]$$

$$= \frac{1}{2}\left[1 - \frac{1}{4} \right]$$

$$= \frac{1}{2}\left(\frac{3}{4} \right)$$

$$= \frac{3}{8}$$

21.   (C)

$$\lim_{x \to 0} \frac{x^{-3/2}}{\dfrac{2}{x^2} + \dfrac{105}{x}} \boxed{\dfrac{x^2}{x^2}} = \lim_{x \to 0} \frac{3}{2 + 105x}$$

$$= \frac{3}{2+0}$$

$$= \frac{3}{2}$$

22.   (D)

Consider $\dfrac{d}{dx} \int f(x)\, dx = f(x)$. So the following graph is the graph of the derivative of $\int f(x)\, dx$. Note the graph of the derivative indicates:

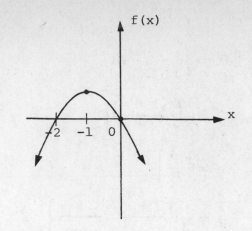

interval	$f(x)$	$f(x)\, dx$
$(-\infty, -2)$	negative	function is decreasing
$(-2, 0)$	positive	function is increasing
$(0, +\infty)$	negative	function is decreasing

Therefore the graph of the function has a relative minimum when $x = -2$ and a relative maximum when $x = 0$. Choice (D) is the only choice with correct locations of the relative minimum and relative maximum. Since $\int f(x)\, dx = F(x) + C$ for any real number C, the graph would be adjusted up if $C > 0$ and down if $C < 0$.

23. (C)

If $f(x) = \int (1-2x)^3\, dx$ then $f(x) = \dfrac{d}{dx} \int (1-2x)^3\, dx$

So $f'(x) = (1-2x)^3$ since $\dfrac{d}{dx} \int f(x)\, dx = f(x)$ by definition

$f''(x) = 3(1-2x)^2 (-2) = -6(1-2x)^2$

$f''\left(\dfrac{1}{2}\right) = -6\left(1 - 2\left(\dfrac{1}{2}\right)\right)^2 = 0$

24. (C)

$F(x) = \displaystyle\int_1^x f(t)\,dt$ and the graph of $f(t)$ is as shown.

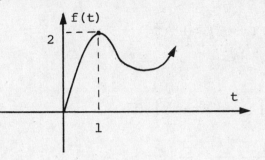

Now $F'(x) = f(x)$ using the Fundamental Theorem of Calculus.

So $F'(1) = f(1) = 2$, using the graph of f.

25. (E)

$$\int x^{-1}\,dx = \int \frac{1}{x}\,dx$$

$$= \ln x + C \ \ \text{since} \ \frac{d}{dx}(\ln x + C) = \frac{1}{x}$$

26. (C)

Solve the equations simultaneously to find the points of intersection.

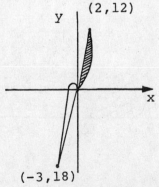

135

$$x^3 + x^2 = 6x$$

$$x^3 + x^2 - 6x = 0$$

$$x(x+3)(x-2) = 0$$

$$x = 0 \quad \text{or} \quad x = -3 \quad \text{or} \quad x = 2$$

$$A = \int_0^2 [6x - (x^3 + x^2)] \, dx$$

$$= \left(\frac{6x^2}{2} - \frac{x^4}{4} - \frac{x^3}{3} \right) \Big|_0^2$$

$$= \frac{16}{3}$$

27.   (B)

$$\lim_{x \to 0} \frac{\arctan x}{\tan x} = \lim_{x \to 0} \frac{\dfrac{1}{1+x^2}}{\sec^2 x}$$

$$= \frac{1}{1}$$

$$= 1 \qquad \text{using L'Hôpital's Rule.}$$

28.   (B)

The period of $y = \sin(ax)$ is $\dfrac{2\pi}{a}$.

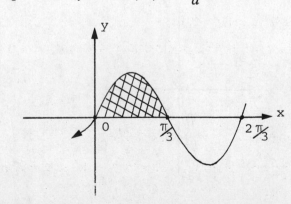

136

$$\int_0^{\pi/3} \sin(3x)\, dx = \frac{1}{3} \int_0^{\pi} \sin u\ du \quad \text{where} \quad u = 3x$$

$$A = \frac{1}{3}\left(-\cos u\right)\Big|_0^{\pi}$$

$$= \frac{1}{3}\left[-(-1) - (-1)\right]$$

$$= \frac{2}{3}$$

29.   (E)

The graph of $V(t)$ is

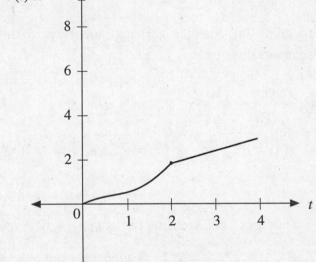

The distance travelled between $t = 1$ and $t = 2$ is

$fnInt(x^\wedge 2, x, 1, 2) = 2.3$.

The distance travelled between $t = 2$ and $t = 3$ is

$fnInt(x + 2, x, 2, 3) = 4.5$.

Hence, from $t = 1$ to $t = 3$, the particle travelled $2.3 + 4.5 = 6.8$.

30.   (C)

Use your calculator to solve this problem:

$fnInt\left(x + \dfrac{1}{x^\wedge 1.6}, x, 0.1, 2\right)$, which should give 7.53.

137

31. **(C)**

$$\lim_{x \to -4^+} \frac{4x - 6}{2x^2 + 5x - 12} = \lim_{x \to -4^+} \frac{2(2x - 3)}{(2x - 3)(x + 4)}$$

$$= \lim_{x \to -4^+} \frac{2}{x + 4}$$

$$= +\infty$$

Note: If $\lim_{x \to a} f(x) = c$ and $\lim_{x \to a} g(x) = 0$,

then $\lim_{x \to a} \dfrac{f(x)}{g(x)} = \pm \infty$

depending on whether $g(x)$ approaches zero through positive or negative values of x.

32. **(E)**

$$g(x) = e^{-x^2} \quad \text{so} \quad g'(x) = -2x \, e^{-x^2} \Rightarrow g'(x) > 0$$

for $x < 0$

Thus $g(x)$ is increasing for x in $(-\infty, 0)$

$$g(-x) = e^{-(-x)^2} = e^{-x^2} \Rightarrow g(x) \text{ is even .}$$

$g(x)$ is symmetric with respect to the y-axis, but is not symmetric with respect to the x-axis because $y = e^{-x^2}$ is not the same graph as $-y = e^{-x^2}$. $g''(x) = 2e^{-x^2}(2x^2 - 1)$ so points of inflection occur when $2x^2 - 1 = 0 \Rightarrow x = \sqrt{1/2}$. Therefore, none of the responses are correct.

33. **(C)**

The left-hand derivative at $x = 0$ is negative and the right-hand derivative at $x = 0$ is positive, so the function is not differentiable at $x = 0$. Therefore, f is not differentiable on the interval $(-2, 2)$.

34. (C)

Let $u = \sec x + \tan x$ and $du = (\sec x \tan x + \sec^2 x)\, dx$

So $\int \frac{1}{u}\, du = \int \frac{(\sec x + \tan x)\,\sec x}{(\tan x + \sec x)}\, dx$

$$= \int \sec x\, dx$$

But $\int \frac{1}{u}\, du = \ln |u| + C$

$$= \ln |\sec x + \tan x| + C \quad \text{which is (A)}$$

(B) follows since

$$- \ln |\sec x - \tan x| = \ln \left| \frac{1}{\sec x - \tan x} \right|$$

$$= \ln |\sec x + \tan x|$$

(D) is correct because

$$\sec x + \tan x = \frac{1}{\cos x} + \frac{\sin x}{\cos x}$$

$$= \frac{1 + \sin x}{\cos x}$$

(E) is correct because

$$\frac{1 + \sin x}{\cos x} = \frac{\cos x}{1 - \sin x}$$

35. (A)

Let $f(x) = 8 + 2x - x^2$ and $g(x) = 6x^3$. Draw the graph of $f(x)$ and $g(x)$.

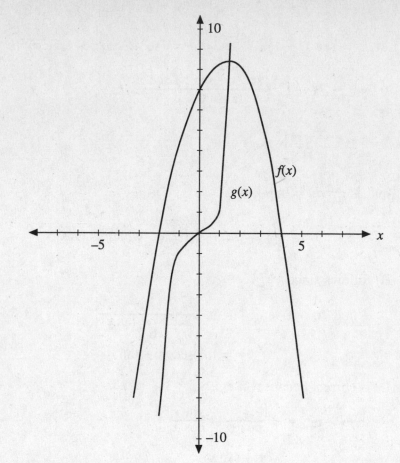

One solution is between 1 and 2 on the *x*-axis. Reset window to [1, 2] · [8, 10]. This solution can be found to be 1.143 .

36.　(C)

Use your calculator to solve this problem. For example,

$$\text{der } 1\left(\frac{\sin x \cos x}{\cos 9x \tan 2x}, x, 0.5\right).$$

–49.5 will be given as the answer.

37.　(A)

Draw the graphs of $f(x)$ and $f'(x)$.

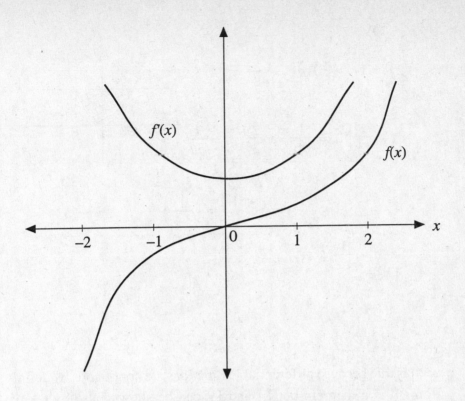

Obviously, the minimum occurs at $x = 0$. Set the window to $[-2, 2] \cdot [-1, 1]$ and trace the x variable to $x = 0$. The minimum is $f'(0) = 0.5$.

38.   (A)

(A) is false because the maximum value of a continuous function may occur at a critical value which is not an endpoint.

(B) is true because of the intermediate value theorem.

(C) is true because of the definition of continuity on [a, b].

(D) is true for any a, b as a consequence of the triangle inequality, i.e., $|a| - |b| \leq |a - b|$.

(E) is true because of the definition of continuity.

39.   (D)

$$2\int_0^{\sqrt{c}} (c - x^2)\ dx = \frac{1}{48}$$

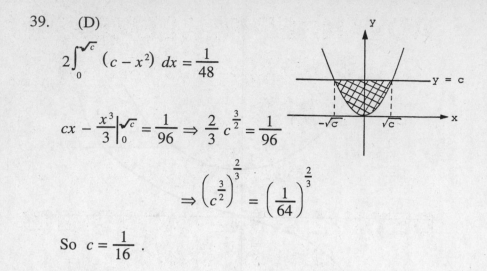

$$cx - \frac{x^3}{3}\Big|_0^{\sqrt{c}} = \frac{1}{96} \Rightarrow \frac{2}{3}c^{\frac{3}{2}} = \frac{1}{96}$$

$$\Rightarrow \left(c^{\frac{3}{2}}\right)^{\frac{2}{3}} = \left(\frac{1}{64}\right)^{\frac{2}{3}}$$

So $c = \frac{1}{16}$.

40.   (C)

I is true because differentiable functions are continuous by definition. II is a theorem but III and IV can be shown false by taking $f(x) = -x$ and $g(x) = x^3$. Note $f'(x) = -1$ and $g'(x) = 3x^2$ which satisfy all the initial conditions. However, $f(x) > g(x)$ for $x < 0$ so III is false and similarly for IV.

ADVANCED PLACEMENT CALCULUS AB EXAM III

SECTION II

DETAILED EXPLANATIONS OF ANSWERS

1.   (A)

Note that the only intercept is at the origin (0,0) since 0 is the only number to make the numerator zero. The denominator is zero when $x = 1$, so sketch in a vertical asymptote at $x = 1$. The limit of $f(x)$ as $x \to +\infty$ is zero through positive values of $f(x)$. The limit $f(x)$ as $x \to +\infty$ is zero through negative values of $f(x)$. Sketch a portion of the graph as indicated in figure (1a). Next calculate where the first and second derivatives change sign.

$$f'(x) = \frac{\left[(x-1)^2 \cdot 1\right] - \left[x(2)(x-1)\right]}{(x-1)^4} = \frac{-x-1}{(x-1)^3}$$

(using the quotient rule for derivatives)

```
+ + + + 0 - - - - - - - - - - (-x - 1) ◄— numerator
- - - - - - - - - - 0 + + + + + (x - 1)³ ◄— denominator
   |    |   ⩗   |   ⩗   |    |
  -3   -2  -1   0   1   2    3
decreasing )(incr.   )(decreasing
```

143

Interval	f' numerator	f' denominator	$f'(x)$	Behavior of f
$(-\infty, -1)$	positive	negative	negative	decreasing
$(-1, 1)$	negative	negative	positive	increasing
$(1, +\infty)$	negative	positive	negative	decreasing

$$f''(x) = \frac{(x-1)^3(-1) - (-x-1)^3(x-1)^2}{(x-1)^4} = \frac{2x+4}{(x-1)^4}$$

```
- - - - 0 + + + + + + + + + + + + + (2x + 4)  ← numerator
+ + + + + + + + + + + 0 + + + + + + (x - 1)⁴  ← denominator
     ✳         ✳
 -3  -2  -1   0   1   2   3
concave down )( concave up )( concave up
```

Using the first derivative test, we see that $f'(x) = 0$ when

$$0 = \frac{-x-1}{(x-1)^3} \Rightarrow 0 = -x-1 \Rightarrow x = -1$$

is a critical point. Taking the second derivative, we have

$$f''(x) = \frac{-(x-1)^3 - (-x-1)3(x-1)^2}{(x-1)^6}$$

$$= \frac{(x-1)^2(-x+1+3x+3)}{(x-1)^6} = \frac{(2x+4)}{(x-1)^4}.$$

At $x = -1$, $f''(x) = \frac{2(-1)+4}{(-2)^4} = \frac{2}{16} > 0$. So $(-1, f(-1))$ is a

relative minimum. To find inflection points, we set $f''(x) = 0$.

$$0 = \frac{2x+4}{(x-1)^4} \Rightarrow 2x+4 = 0 \Rightarrow x = -2, \quad \text{So } (-1, f(-1)) \text{ is a}$$

point of inflection.

The completed sketch of the graph is as follows:

144

(B)

f is increasing on $(-1, 1)$ and decreasing on $(-\infty, -1)$ and $(1, \infty)$. f is concave down on $(-\infty, -2)$ and concave up on the intervals $(-2, 1)$ and $(1, +\infty)$. See part (A) for appropriate work.

(C)

The domain of f is all reals except one. The range of f is all reals greater than or equal to $-\dfrac{1}{4}$. The function f does not have an inverse over the domain of f because f is not a one-to-one function. For example, $f\left(-\dfrac{1}{2}\right) = f(-2) = -\dfrac{2}{9}$, so two different values of x can go to the same value of $y = f(x)$.

(D)

Use the point-slope form for the equation of a line:

$$y - y_1 = m(x - x_1), \quad \text{where} \quad x_1 = -2 \quad \text{and} \quad x_2 = -\dfrac{2}{9}$$

and $m = f'(-2) = \dfrac{2 - 1}{(-3)^3}$

$$y - \left(-\dfrac{2}{9}\right) = -\dfrac{1}{27}(x - (-2)) \quad \text{or}$$

$$x + 27y + 8 = 0$$

2.   (A)

Using the first derivative formation,

$$\text{der } 1\left(\dfrac{3}{2 + 105\,x}, x, 0\right)$$

$f(0) = -78.75$ can be found.

(B)

Draw the graph of $f(x)$ for $-1 < x < 1$.

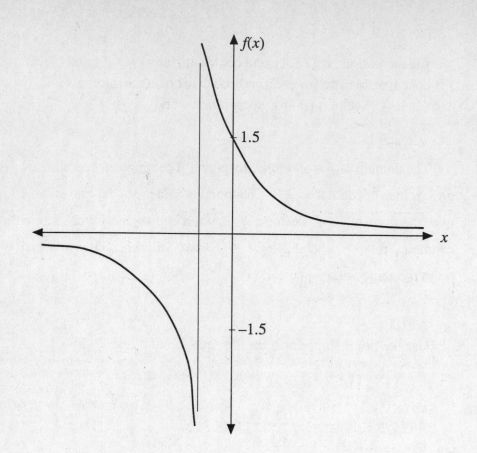

At $x = -0.019$, $f(x)$ jumps from an extremely large negative value to an extremely large positive value. So, $f(x)$ at $x = -0.019$ has no value.

(C)
Using the calculator to $\int_0^2 f(x)$ by

$$fnInt\left(\frac{3}{2+105\,x},\, x, 0, 2\right),$$

you get

$$\int_0^2 f(x)\ dx = 0.1332.$$

3.   (A)
Let t = time in seconds for plant growth.
    x = weight (in grams) of plant culture X at time t.

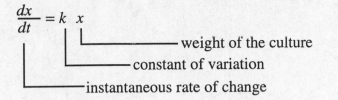

$$\frac{dx}{dt} = k \; x$$

— weight of the culture
— constant of variation
— instantaneous rate of change

    (B)
The solution for the differential equation $f'(t) = k f(t)$ is $f(t) = A e^{kt}$, so let $x = f(t)$ and then $x = A e^{kt}$.

    (C)
For $x = f(t) = A e^{kt}$, let $t = 0$ and then $10 = f(0) = A e^{0}$; so $A = 10$. Let $t = 1$ and then $100 = f(1) = 10e^{k}$.

Dividing by 10 gives $10 = e^{k}$, so $k = \ln 10$. This means $x = f(t) = 10 e^{t \ln 10} = 10 e^{\ln 10t} = 10(10^{t}) = 10^{t+1}$.

If $x = 0.5$ then $f(.5) = 10^{1.5} = 10 \sqrt{10}$ grams.

    (D)
The instantaneous rate of growth is $f'(t) = D_t (10^{t+1})$.

Now $f'(t) = 10^{t+1} \ln 10$, so $f'(1) = 100 \ln 10$ gm/sec.

Average rate of growth $\dfrac{f(1) - f(0)}{1 - 0} = 100 - 10 = 90$ gm/sec.

Therefore $f'(1) -$ Avg. rate $= (100 \ln 10 - 90)$ gm/sec.

4.　(A)

The period for $y = \cos(au)$ is $\dfrac{2\pi}{a}$, so the period for
$y = \cos(2u)$ is π. The first positive x-intercept will be when
$u = \dfrac{\pi}{4}$ so $x^2 = u = \dfrac{\pi}{4}$; therefore, $x = \sqrt{\dfrac{\pi}{4}}$.

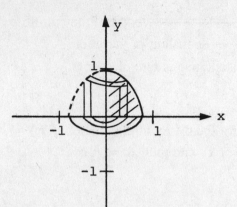

Using the method of shells, the volume $= 2\pi \displaystyle\int_a^b x\, f(x)\, dx$.

$$\text{Volume} = 2\pi \int_0^{\sqrt{\frac{\pi}{4}}} x \cos(2x^2)\, dx = 2\pi \int_0^{\frac{\pi}{2}} \frac{1}{4} \cos(u)\, du$$

where we have set $u = 2x^2$ so $du = 4x\, dx$

$$\text{Volume} = \frac{\pi}{2} \sin u \Big|_0^{\frac{\pi}{2}}$$

$$= \frac{\pi}{2}$$

(B)

$$f'(x) = -\sin(2x^2)\, D_x(2x^2) = -4x\sin(2x^2)$$

$$\text{Average value} = \frac{1}{b-a}\int_a^b f'(x)\, dx$$

$$= \frac{1}{\sqrt{\pi/4}} \int_0^{\sqrt{\frac{\pi}{4}}} -4x \sin(2x^2)\, dx$$

148

Let $u = 2x^2$ and $du = 4x\ dx$ to get

$$-\sqrt{\frac{4}{\pi}} \int_0^{\frac{\pi}{2}} \sin u\ du = \frac{2}{\sqrt{\pi}} \cos u \Big|_0^{\frac{\pi}{2}}$$

$$= -\frac{2\sqrt{\pi}}{\pi}$$

(C)

Using the method of cylindrical washers, the volume is

$$\int_a^b \left\{ [f(x)]^2 - [g(x)]^2 \right\} dx = \int_0^{\sqrt{\frac{\pi}{4}}} \left[1^2 - \left(\cos(2x^2)\right)^2 \right] dx$$

$$= \int_0^{\sqrt{\frac{\pi}{4}}} \sin^2(2x^2)\ dx$$

5. (A)

$$V'(t) = V(t) = \int V'(t)\ dt$$

$$= \int 3\ dt$$

$$= 3t + C$$

Since $V(0) = 0$, $V(t) = 3t$ so $t_0 = \frac{1}{3}V(t_0)$.

Now the volume of a cube V is x^3, the surface area S is $6x^2$, and $\frac{dV}{dt} = \frac{dS}{dt}$ when $t = t_0$.

$$\frac{dV}{dt} = \frac{dV}{dx}\frac{dx}{dt} = \frac{dS}{dx}\frac{dx}{dt} \Rightarrow 3x^2\frac{dx}{dt} = 12x\frac{dx}{dt}$$

$$\Rightarrow 3x^2 = 12x \Rightarrow x = 0, 4.$$

$x = 0$ makes no sense geometrically.
If $x = 4$, $V = x^3 = 64 = 3t_0$.

Therefore, $t_0 = \frac{1}{3} \cdot 64 = 21\frac{1}{3}$ seconds.

(B)

$$\frac{dV}{dt} = 3x^2\frac{dx}{dt} = 3,$$

so $\dfrac{dx}{dt} = \dfrac{1}{x^2} = \dfrac{1}{16}$ cm/sec

(C)
$\dfrac{dS}{dt} = \dfrac{dV}{dt}\dfrac{dS}{dV}$. To find $\dfrac{dS}{dV}$, we note that

$$S = 6x^2 = 6 \cdot \left(x^3\right)^{2/3} = 6V^{2/3}$$

So $\dfrac{dS}{dV} = \dfrac{2}{3} \cdot 6 \cdot V^{-1/3} = \dfrac{4}{V^{1/3}}$.

When $t = t_0$, $V = 64$ (as in part (A)), so $V^{1/3} = 4$

and $\dfrac{dS}{dt} = \dfrac{dV}{dt}\dfrac{dS}{dV} = 3 \cdot \dfrac{4}{4} = 3$ cm^2/sec.

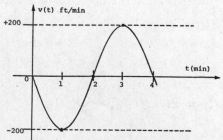

150

6. (A)

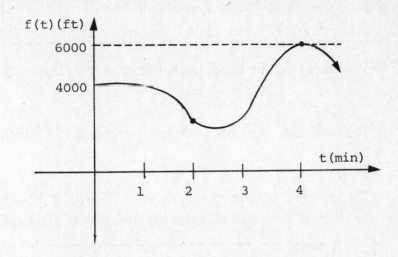

Use the graph of $y = v(t) = f'(t)$ to determine where f is increasing and decreasing as follows:

interval	$v(t) = f'(t)$	behavior of $y = f(t)$
(0, 2)	negative	decreasing in (0, 2)
(2, 4)	positive	increasing in (2, 4)

Note that the graph starts at 4,000 feet and ends at 6,000 feet.

(B)

$f(t)$ is continuous because it is differentiable, as indicated by the given graph of $f'(t)$. If a function is continuous on a closed interval, then the function attains a maximum and minimum value at least once on the interval. The maximum and minimum value of the function on the interval occur at an endpoint or at a critical value. So evaluate the function at its endpoints and critical values.

$f(0) = 4,000$ feet
$f(2) < 4,000$ feet since f is decreasing in $(0, 2)$
$f(4) = 6,000$ feet

Therefore the altitude is a minimum when $t = 2$ seconds.

(C)

The maximum altitude is 6,000 feet. See part (B) for justification.

(D)

The slopes of the tangent lines to $y = v(t)$ will be the values for $y = v'(t) = f''(t)$ which indicates the concavity of $f(t)$ as follows:

interval	slopes of $v(t)$	concavity
(0, 1)	negative	down
(1, 3)	positive	up
(3, 4)	negative	down

THE ADVANCED PLACEMENT EXAMINATION IN

CALCULUS AB

TEST IV

ADVANCED PLACEMENT CALCULUS AB EXAM IV

SECTION I

PART A

Time: 45 minutes
        25 questions

DIRECTIONS: Each of the following problems is followed by five choices. Solve each problem, select the best choice, and blacken the correct space on your answer sheet. Calculators may not be used for this section of the exam.

NOTE:
Unless otherwise specified, the domain of function f is assumed to be the set of all real numbers x for which $f(x)$ is a real number.

1.   Which of the following represents a function?

(A) $x^2 + y^2 = 1$ (C) $x^2 - y = 0$

(B) $y^2 = x$ (D) $y = \pm\sqrt{1 - x^2}$

(E)

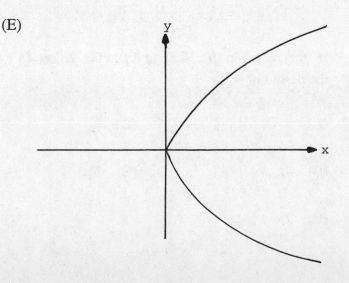

2. If $f(x) = x - 2$ and $g(x) = |x|$, then

(A) $(g \circ f)(3) + f(3) = 0$

(B) $(g \circ f)(1) - f(1) = 0$

(C) $\dfrac{(g \circ f)(0)}{f(0)} = -1$

(D) $\dfrac{(g \circ f)(0)}{f(0)} = 1$

(E) $2(g \circ f)(x) \neq f(x)$ for every x

3. If $f(x) = \dfrac{x^2}{1 - x^2}$, then

(A) Domain of $f = R\backslash\{1\}$

(B) f is an odd function

(C) The line $y = -1$ is a horizontal asymptote

(D) $x = -1$ is the only vertical asymptote

(E) f is never zero

4. If $f(x) = a \sin(bx + c)$ and $f'(x) = 2a \cos(bx + c)$, then the period of f is

(A) π (D) $\dfrac{\pi}{2}$

(B) 4π (E) 3π

(C) 2π

5. If the graph of f is as shown below, then

$$\lim_{x \to 3} \left([f(x)]^2 + 2f(x) + 1 \right) =$$

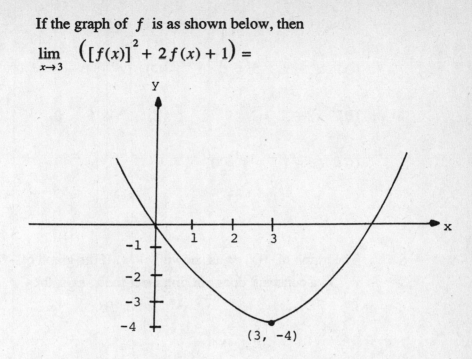

(3, −4)

(A) 8

(D) 9

(B) 10

(E) 25

(C) 11

6. If $\sin\left(\frac{1}{x}\right) \le f(x) \le \cos\left(\frac{1}{x}\right)$, $x \ne 0$, then

$$\lim_{x \to 0} x^2 f(x) =$$

(A) 1

(D) ∞

(B) −1

(E) 0

(C) Does not exist

157

7. Which of the following has a real root?

(A) $x^3 - x + 5 = 0$ (D) $(x - 1)^2 + 3 = 0$

(B) $x^2 + 1 = 0$ (E) $x^2 + 5 = 0$

(C) $(x + 1)^2 + 1 = 0$

8. The graph of $f(x)$ is as shown below. If the graph of $-f(x) + c$, c a constant, does not intersect the x-axis, then c must be

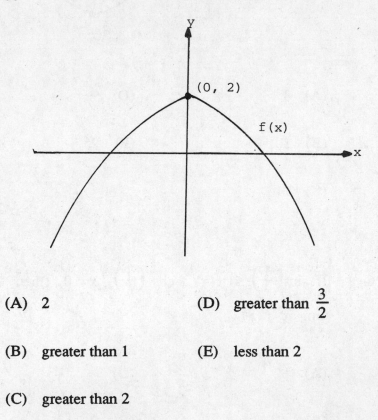

(A) 2 (D) greater than $\dfrac{3}{2}$

(B) greater than 1 (E) less than 2

(C) greater than 2

9. If f has a derivative at $x = 2$ and $g(x) = (x - 2)^2 f(x)$, then

(A) $g'(2)$ does not exist

(D) $g'(2) < 0$

(B) $g'(2) = 0$

(E) $g'(2) > 0$

(C) $g'(2)$ cannot be determined

10. If $f(x) = \sqrt{x + 2}$, then

$$\lim_{h \to 0} \frac{f(2 + h) - f(2)}{h} =$$

(A) 4

(D) $\frac{1}{4}$

(B) 0

(E) 1

(C) $\frac{1}{2}$

11. Given $f(x) = 2x^2 - 3x + 5$ and $g(x) = x^2 + 2x + 4$, if $f'(a) = g'(a)$ then $a =$

(A) 2

(D) $\frac{2}{5}$

(B) $\frac{5}{6}$

(E) $\frac{5}{2}$

(C) $-\frac{2}{5}$

12. If $y = A \sin x + B \cos x$, then

(A) $y + \dfrac{d^2y}{dx^2} = 0$ (D) $y + \dfrac{d^2y}{dx^2} < 0$

(B) $y + \dfrac{d^2y}{dx^2} \neq 0$ (E) $-y + \dfrac{d^2y}{dx^2} = 0$

(C) $y + \dfrac{d^2y}{dx^2} > 0$

13. Let $f(x) = \dfrac{2}{3}x^{\frac{3}{2}}$ and suppose that the line $y = 2.5x$ is tangent to $f(x)$ at x_0. x_0 must be approximately equal:

(A)   −6.25   (D)   0.71

(B)   4.00   (E)   12.50

(C)   6.25

14. The value of $\dfrac{dy}{dx}$ when $x = 1$ and $y = -1$ given that $4x^2 + 2xy - xy^3 = 0$ is

(A) 7   (D) 11

(B) 5   (E) 6

(C) 9

160

15. The domain of $y = \sqrt{(x-1)(x-2)}$ is

(A) $|x| < 2$

(D) $(-\infty, 1] \cup [2, \infty)$

(B) $(1, 2)$

(E) $[1, 2]$

(C) $|x| > 1$

16. If the graph of $f(x)$ is as shown below, then

(A) Domain of $\dfrac{1}{f(x)} = (1, 4)$

(B) Domain of $\dfrac{1}{f(x)} = [1, 4]$

(C) $f'(x) > 0$ on $(1, 4)$

(D) $f'(x) < 0$ on $(1, 4)$

(E) $f'(x_0) = 0$ at some point in $(1, 4)$

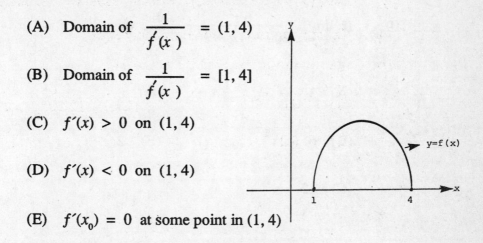

17. If $\dfrac{dy}{dx} = 3 \cos^2 x - 3 \sin^2 x$, then $y =$

(A) $\sin(2x) + C$

(D) $\dfrac{3}{2} \cos(2x) + C$

(B) $3 \sin(2x) + C$

(E) $\dfrac{2}{3} \sin(2x) + C$

(C) $\dfrac{3}{2} \sin(2x) + C$

18. $\lim\limits_{x \to 0} \dfrac{x - \tan x}{x - \sin x} =$

(A) –2

(D) $-\dfrac{1}{2}$

(B) 2

(E) Does not exist

(C) 0

19. If $\lim\limits_{n \to \infty} \dfrac{6n^2}{200 - 4n + kn^2} = \dfrac{1}{2}$, then $k =$

(A) 3

(D) 8

(B) 6

(E) 2

(C) 12

20. If $\lim\limits_{n \to \infty} \left(1 + \dfrac{1}{n}\right)^{kn} = \dfrac{1}{e^2}$, then $k =$

(A) 1

(D) $-\dfrac{1}{e}$

(B) –2

(E) $\dfrac{1}{e}$

(C) $\dfrac{1}{2}$

21. If $y = -\dfrac{1}{\sqrt{x^2 + 1}}$, then $\dfrac{dy}{dx} =$

(A) $\dfrac{x}{(x^2 + 1)^{1/2}}$

(D) $\dfrac{x}{(x^2 + 1)^{3/2}}$

(B) $-\dfrac{x}{(x^2 + 1)^{1/2}}$

(E) $\dfrac{x}{x^2 + 1}$

(C) $-\dfrac{x}{(x^2 + 1)^{3/2}}$

22. If $y = \ln[(x + 1)(x + 2)]$, then $\dfrac{dy}{dx} =$

(A) $\dfrac{1}{x + 1} + (x + 2)$

(B) $\dfrac{1}{(x + 2)} + (x + 1)$

(C) $\dfrac{1}{(x + 1)(x + 2)}$

(D) $\dfrac{x + 1}{x + 2}$

(E) $\dfrac{1}{x + 1} + \dfrac{1}{x + 2}$

23. If $f(x) = ae^{kx}$ and $\dfrac{f'(x)}{f(x)} = -\dfrac{5}{2}$ then $k =$

(A) -5

(D) $\dfrac{5}{2}$

(B) $-\dfrac{5}{2}$

(E) $\dfrac{2}{5}$

(C) $-\dfrac{2}{5}$

24. The rate of change of the area of an equilateral triangle with respect to its side S at $S = 2$ is approximately:

(A)   0.43

(D)   7.00

(B)   1.73

(E)   0.50

(C)   0.87

25. If $f(x) = e^{\frac{x^3}{3} - x}$ then $f(x)$

(A) increases in the interval $(-1, 1)$

(B) decreases for $|x| > 1$

(C) increases in the interval $(-1, 1)$ and decreases in the invervals $(-\infty, -1) \cup (1, \infty)$

(D) increases in the intervals $(-\infty, -1) \cup (1, \infty)$ and decreases in the inverval $(-1, 1)$

(E) increases in the inverval $(-\infty, \infty)$

SECTION I

PART B

Time: 45 minutes
        15 questions

DIRECTIONS: Calculators may be used for this section of the test. Each of the following problems is followed by five choices. Solve each problem, select the best choice, and blacken the correct space on your answer sheet.

NOTES:

1. Unless otherwise specified, answers can be given in unsimplified form.

2. The domain of function f is assumed to be the set of all real numbers x for which $f(x)$ is a real number.

26.   If $f(x) = x - \dfrac{1}{x}$, which of the following is NOT true?

(A) $f\left(\dfrac{1}{x}\right) = -f(x)$ (D) $f(-x) = -f(x)$

(B) $f\left(\dfrac{1}{x}\right) = f(-x)$ (E) None of these

(C) $f(x) = f(-x)$

27.   If x and y are two positive numbers such that $x + y = 20$ and xy is as large as possible, then

(A) $x = 12$ and $y = 8$ (D) $x = 20$ and $y = 0$

(B) $x = 10$ and $y = 10$ (E) $x = 8$ and $y = 12$

(C) $x = 5$ and $y = 15$

165

28. If the rate of change of $f(x)$ at $x = x_0$ is twice the rate of change of $f(x)$ at $x = 4$, and $f(x) = 2\sqrt{x}$, then x_0 is

(A) 8

(D) 16

(B) 1

(E) 4

(C) 2

29. The area enclosed by $f(x) = x^3 + x^2$ and $g(x) = \ln(x + 1)$ for $x > 0$ is

(A) 0.0513

(D) 2.89

(B) 0.01

(E) 7.8

(C) 2.5

30. The area in the first quadrant that is enclosed by $y = \sin x \cos x$ and the x-axis from $x = 0$ to the first x-intercept on the positive side is

(A) 1

(D) 2.5

(B) 0.56

(E) 1.5

(C) 0.78

31.   If the graph of f is as shown below, then the graph of $y = 2 + f(x)$

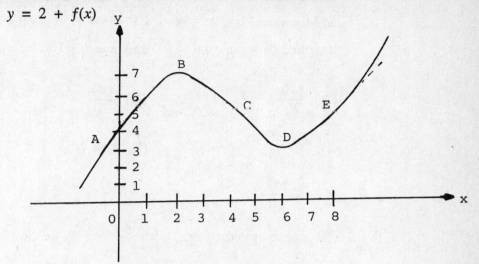

(A)   is concave downward in the interval $(2, 6)$

(B)   is concave upward in the interval $(0, 8)$

(C)   has an inflection point at $x = 5$

(D)   is concave downward in the interval $(0, 3)$

(E)   is concave downward in the interval $(0, 8)$

32.   If $f(x) = g(x) - \dfrac{1}{g(x)}$, $g(0) = 3$ and $g'(0) = 2$, then $f'(0)$ is approximately:

(A)   1.778   (D)   2.222

(B)   2.333   (E)   1.222

(C)   3.222

33. If the acceleration (ft/sec²) of a moving body is $\sqrt{4t+1}$ and the velocity at $t = 0$, $v(0) = -4\frac{1}{3}$, then the distance travelled between time $t = 0$ and $t = 2$ is

(A) $\dfrac{119}{30}$

(B) $-\dfrac{119}{30}$

(C) $-\dfrac{149}{30}$

(D) $\dfrac{149}{30}$

(E) -4

34. If the volume of a cube is increasing at a rate of 300 in³/min at the instant when the edge is 20 inches, then the rate at which the edge is changing is

(A) $\dfrac{1}{4}$ in/min

(B) $\dfrac{1}{2}$ in/min

(C) $\dfrac{1}{3}$ in/min

(D) 1 in/min

(E) $\dfrac{3}{4}$ in/min

35. $6\displaystyle\int_{-1}^{1} \dfrac{x^3 + 1}{x + 1}\, dx =$

(A) 4

(B) 2

(C) 12

(D) 8

(E) 16

36. If $f(x) = x^3 - 2x$, then f has

(A) a relative Max. at $x = \sqrt{\frac{2}{3}}$ and a relative Min. at $x = -\sqrt{\frac{2}{3}}$

(B) an absolute Max. at $x = -\sqrt{\frac{2}{3}}$

(C) a relative Min. at $x = \sqrt{\frac{2}{3}}$ and a relative Max. at $x = -\sqrt{\frac{2}{3}}$

(D) an absolute Min. at $x = \sqrt{\frac{2}{3}}$

(E) a relative Min. at $x = -\sqrt{\frac{2}{3}}$ and a relative Max. at $x = \sqrt{\frac{2}{3}}$

37. If $f(x) = 256\, x^{-\frac{1}{2}} + 64\, x^{\frac{1}{2}} + 3x^{\frac{2}{3}}$, then $f'(64)$ equals

(A) 4.25   (D) 10.25

(B) 8.75   (E) 5.78

(C) 0.75

38. The smallest value of $y = x^2\,(1 - x^{-1})$ is

(A) 0   (D) –1

(B) –0.25   (E) 1

(C) 0.25

39. Given the following graph of the continuous function $f(x)$,

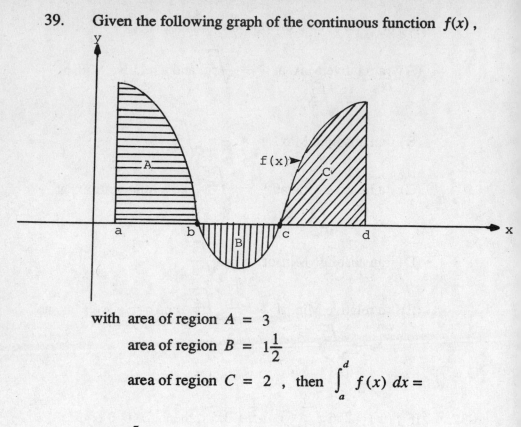

with area of region A = 3

area of region B = $1\frac{1}{2}$

area of region C = 2 , then $\int_a^d f(x)\,dx$ =

(A) $\dfrac{5}{2}$ (D) 5

(B) $\dfrac{7}{2}$ (E) $\dfrac{2}{5}$

(C) $\dfrac{13}{2}$

40. $\int x\cos\,(4x)\,dx$ =

(A) $\dfrac{1}{4}\sin\,(4x) + C$

(B) $\dfrac{x}{4}\sin\,(4x) + C$

(C) $\dfrac{x}{4}\sin\,(4x) + \dfrac{1}{16}\cos\,(4x) + C$

(D) $\dfrac{x}{4} \sin (4x) - \dfrac{1}{16} \cos (4x) + C$

(E) $\dfrac{x}{4} - \dfrac{1}{16} \cos (4x) + C$

ADVANCED PLACEMENT CALCULUS AB EXAM IV

SECTION II

Time: 1 hour and 30 minutes
        6 problems

DIRECTIONS: Show all your work. Grading is based on the methods used to solve the problem as well as the accuracy of your final answers. Please make sure all procedures are clearly shown.

NOTES:
1. Unless otherwise specified, answers can be given in unsimplified form.

2. The domain of function f is assumed to be the set of all real numbers x for which $f(x)$ is a real number.

1. If $f(x) = \dfrac{1}{2} + \dfrac{1}{2} \cos(2x)$, $0 \le x \le \pi$, then

   (A) find $\displaystyle \lim_{x \to \frac{\pi}{4}} f(x)$.

   (B) find the average value of f on $[0, \pi]$.

   (C) show that $|f(b) - f(a)| \le |b - a|$ for any $a < b$.

2. Given $f(x) = 2x^2 - 2x + 5$ and $g(x) = x^2 + 2x + 4$

(A) Find the values of x where $f(x) = g(x)$.

(B) If $f'(a) = g'(a)$, find a .

(C) Use the results from (A) to find $\int_a^b (g(x) - f(x))\ dx$, where a and b are where $f(x) = g(x)$.

3. Let f be an even function which has a derivative at every value of x in its domain. If $f(2) = 1$ and $f'(2) = 5$, then

(A) Find $f'(-2)$ and $f'(0)$.

(B) Let L_1 and L_2 be the tangents to the graph of f at $x = 2$ and $x = -2$, respectively. Find the coordinates of the point p at which L_1 and L_2 intersect.

4. The graphs of f, g, and h are as given below:

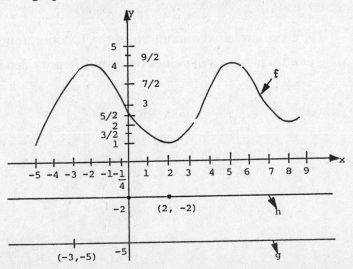

(A) Where is $(h - g)f$ concave upward and where is it concave downward?

(B) Sketch $(h - g)f$.

5.   (A) At which point(s), if any, do $y = \sin x$ and $y = \cos x$ intersect in the interval $\left[0, \dfrac{\pi}{2}\right]$

(B) Find the area of the region between $y = \sin x$, $y = \cos x$, from $x = 0$ to $x = \dfrac{\pi}{2}$.

(C) Set up an integral for the volume obtained by rotating the region in (b) about the x-axis.

6.   (A) Determine the constants a and b in order for the function

$$f(x) = x^3 + ax^2 + bx + c$$

to have a relative minimum at $x = 4$ and a point of inflection at $x = 1$.

(B) Find a relative maximum of the function found in (A) after plugging in values of a and b, provided that $f(0) = 1$.

ADVANCED PLACEMENT
CALCULUS AB
EXAM IV

ANSWER KEY

Section I

1.	C		21.	D
2.	C		22.	E
3.	C		23.	B
4.	A		24.	B
5.	D		25.	D
6.	E		26.	C
7.	A		27.	B
8.	C		28.	B
9.	B		29.	A
10.	D		30.	B
11.	E		31.	D
12.	A		32.	D
13.	C		33.	C
14.	A		34.	A
15.	D		35.	E
16.	E		36.	C
17.	C		37.	A
18.	A		38.	B
19.	C		39.	B
20.	B		40.	C

Section II
See Detailed Explanations of Answers.

ADVANCED PLACEMENT
CALCULUS AB EXAM IV

SECTION I

DETAILED EXPLANATIONS
OF ANSWERS

1.　　(C)

$$x^2 - y = 0 \iff y = x^2$$

is represented by the parabola shown below, and no vertical line intersects the graph in more than one point, so $y = x^2$ is a function.

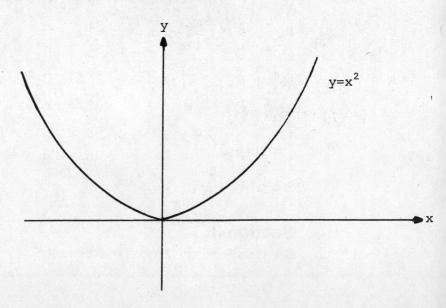

The graph of $y = \pm \sqrt{1 - x^2}$ is also the circle as shown below.

The graph of $y^2 = x$ is the one shown in (E).

In (A), (B), (D) and (E) there is a line parallel to the y-axis (a vertical line) that cuts the graph of the given relations more than once. Hence they do not represent functions.

The graph of $x^2 + y^2 = 1$ is the circle:

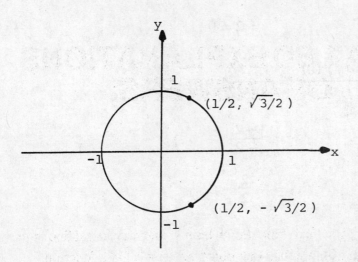

2.　(C)

$$(g \circ f)\,(x) = g(f(x))$$

$$= g(x - 2)$$

$$= |x - 2|$$

Therefore,

(i) $(g \circ f)(3) + f(3) = |3 - 2| + (3 - 2)$

$$= 1 + 1$$

$$= 2 \neq 0 \ .$$

(ii) $(g \circ f)(1) - f(1) = |1 - 2| - (1 - 2)$

$$= |-1| - (-1)$$

$$= 1 + 1$$

$$= 2 \neq 0$$

(iii) $\dfrac{(g \circ f)(0)}{f(0)} = \dfrac{|0 - 2|}{0 - 2}$

$$= \dfrac{|-2|}{-2}$$

$$= -\dfrac{2}{2}$$

$$= -1 \neq 1$$

(iv) $2(g \circ f)(x) = 2|x - 2|$

$$f(x) = x - 2$$

$$\Rightarrow 2(g \circ f)(x) = f(x)$$

for $x = 2$, since both are zero at $x = 2$.
Hence, only (C) is true.

3.   (C)

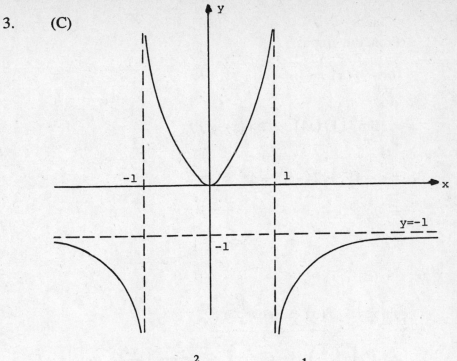

$$f(x) = \frac{x^2}{1-x^2} = -1 + \frac{1}{1-x^2}.$$

The graph of f shows that (D) and (E) are false and (C) is true. Also, since the graph is symmetric with respect to the y-axis, it shows that f is an even function, so (B) is false. Domain $f = R\setminus\{-1, 1\}$, so (A) is false.

4.   (A)

$f(x) = a \sin(bx + c)$

$\Rightarrow f'(x) = ab \cos(bx + c)$ by the chain rule

and $f'(x) = 2a \cos(bx + c)$ from what is given, $\Rightarrow b = 2$.

Hence $f(x) = a \sin(2x + c)$ and the period $= \dfrac{2\pi}{2} = \pi$.

5.   (D)

From the graph,

$$\lim_{x \to 3} f(x) = -4$$

$$\therefore \lim_{x \to 3} \left([f(x)]^2 + 2f(x) + 1 \right)$$

$$= (-4)^2 + 2(-4) + 1$$

$$= 9$$

6.   (E)

$$\sin \frac{1}{x} \le f(x) \le \cos \frac{1}{x}$$

$$\Leftrightarrow x^2 \sin \left(\frac{1}{x} \right) \le x^2 f(x) \le x^2 \cos \left(\frac{1}{x} \right), \quad \text{for } x \ne 0.$$

Furthermore, $\lim_{x \to 0} x^2 \sin \left(\frac{1}{x} \right) = 0$, since $\left| \sin \frac{1}{x} \right| \le 1$.

$$\lim_{x \to 0} x^2 \cos \left(\frac{1}{x} \right) = 0, \quad \text{since } \left| \cos \frac{1}{x} \right| \le 1.$$

Hence $\lim_{x \to 0} x^2 f(x) = 0$, by the limit comparison test.

<u>Note:</u> The sine or cosine of any angle is always between -1 and 1. Therefore, even though $\frac{1}{x}$ gets larger and larger as $x \to 0$ $\sin \left(\frac{1}{x} \right)$ and $\cos \left(\frac{1}{x} \right)$ are confined to lie between -1 and 1.

7.   (A)

Let $f(x) = x^3 - x + 5$

It is possible to find two values of x at which f has different signs, say at $x = -2$ and $x = 0$:

$$f(-2) = (-2)^3 - (-2) + 5$$

$$= -8 + 2 + 5$$

$$= -1 < 0 \text{ and}$$

$$f(0) = 5 > 0$$

Therefore by the Intermediate Value Theorem, $x^3 - x + 5 = 0$ has a real root between $x = -2$ and $x = 0$.

All of the other choices have imaginary roots only, as they are of the form $y^2 = -c$, $c > 0$

For example, in (D) $(x - 1)^2 + 3 = 0 \Rightarrow (x - 1)^2 = -3$ and only an imaginary number can have a square which is a negative number.

8.   (C)

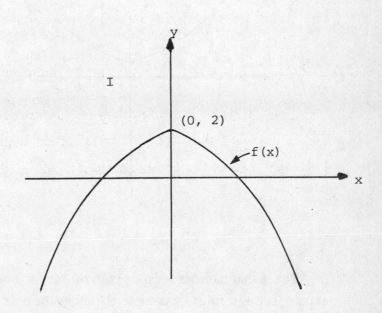

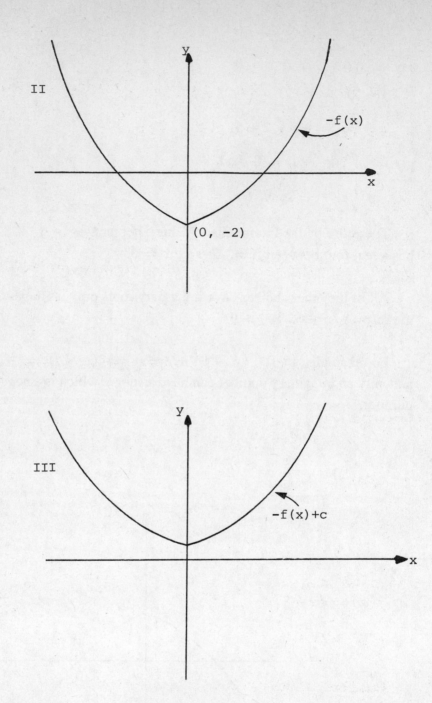

We see that in order for the graph of $f(x)$ to not intersect the x-axis, $-f(x) + c$ must lie completely above the axis. Since the lowest point on the graph $-f(x)$ is the point $(0, -2)$, this point must be raised (along with the rest of the graph) to above the x-axis. This can be accomplished by adding $c > 2$.

9.   (B)

$g'(x) = 2(x-2) f(x) + (x-2)^2 f(x)$, by the product rule.

$\Rightarrow g'(2) = 2(0) f(2) + (2-2)^2 f(2)$

$= 0 + 0 = 0$.

10.   (D)

$\lim_{h \to 0} \dfrac{f(2+h) - f(2)}{h} = f(2)$, by the definition of derivative.

But $f(x) = \dfrac{d}{dx}\left((x+2)^{1/2}\right)$

$= \dfrac{1}{2}(x+2)^{-1/2}$,   by the power rule and chain rule,

$\Rightarrow f(2) = \dfrac{1}{2}(2+2)^{-1/2}$

$= \dfrac{1}{2}(4)^{-1/2}$

$= \dfrac{1}{4}$

11.   (E)

$f(x) = 4x - 3$

$g'(x) = 2x + 2$

Therefore, $f(a) = g'(a)$

$\Rightarrow 4a - 3 = 2a + 2$

$\Rightarrow 2a = 5$

$\Rightarrow a = \dfrac{5}{2}$

12. (A)

$y = A \sin x + B \cos x$

$\Rightarrow \dfrac{dy}{dx} = A \cos x - B \sin x$

$\Rightarrow \dfrac{d^2 y}{dx^2} = \dfrac{d}{dx} (A \cos x - B \sin x)$

$= -A \sin x - B \cos x$

$= -(A \sin x + B \cos x)$

$= -y$

Therefore, $y + \dfrac{d^2 y}{dx^2} = y - y = 0$

13. (C)

$f'(x) = \dfrac{dy}{dx} = \left(\dfrac{3}{2}\right)\left(\dfrac{2}{3}\right) x^{\frac{3}{2}-1} = x^{\frac{1}{2}}$

by power rule for differentiation

$\Rightarrow f'(x_0)$ must equal the slope of the line $y = 2.5x$

$\Rightarrow f'(x_0) = 2.5$

$\Rightarrow x_0^{\frac{1}{2}} = 2.5$

$2.5 \, x^2 = 6.25$

$\Rightarrow x_0 = 6.25$

14. (A)

By implicit differentiation,

$$8x + 2x\frac{dy}{dx} + 2y - y^3 - 3xy^2\frac{dy}{dx} = 0$$

$$\Leftrightarrow (3xy^2 - 2x)\frac{dy}{dx} = 8x + 2y - y^3,$$

after moving the terms containing $\frac{dy}{dx}$ to one side,

$$\Rightarrow \frac{dy}{dx} = \frac{8x + 2y - y^3}{3xy^2 - 2x}.$$

Therefore, when $x = 1$ and $y = -1$,

$$\frac{dy}{dx} = \frac{8 - 2 - (-1)^3}{3 \cdot 1(-1)^2 - 2 \cdot 1}$$

$$= \frac{8 - 2 + 1}{1}$$

$$= 7.$$

15. (D)

	$x < 1$	$x = 1$ ↓ $1 < x < 2$	$x = 2$ ↓ $2 < x$
$(x - 1)$	Negative	0 Positive	Positive
$(x - 2)$	Negative	Negative 0	Positive
$(x - 1)(x - 2)$	Positive	0 Negative	Positive

As the chart shows, when $1 < x < 2$, $(x - 1)(x - 2) < 0$

$\Rightarrow y = \sqrt{(x - 1)(x - 2)}$ is undefined. Otherwise, y is defined.

16. (E)

Since $f(1) = 0$ and $f(4) = 0$, by the mean value theorem (or by Rolle's theorem), there is an $x_0 \in (1, 4)$ such that

$$f'(x_0) = \frac{f(4) - f(1)}{4 - 1} = 0$$

17. (C)

$$\frac{dy}{dx} = 3\cos^2 x - 3\sin^2 x$$

$$= 3\left(\cos^2 x - \sin^2 x\right)$$

$$= 3\cos(2x),$$

since $\cos(2x) = \cos^2 x - \sin^2 x$

$$\Rightarrow y = \int 3\cos(2x)\,dx$$

By the method of substitution, if we let

$$2x = u$$

$$\Rightarrow 2dx = du$$

$$\Rightarrow dx = \frac{du}{2}$$

$$\Rightarrow \int 3\cos(2x)\,dx = 3\int \cos u \left(\frac{du}{2}\right)$$

$$= \frac{3}{2}\sin u + c, \quad \text{where } u = 2x$$

$$= \frac{3}{2}\sin(2x) + c.$$

18. (A)

$\dfrac{x - \tan x}{x - \sin x}$ is an indeterminate form of the type $\dfrac{0}{0}$.

Also: $\dfrac{(x - \tan x)'}{(x - \sin x)'} = \dfrac{1 - \sec^2 x}{1 - \cos x}$ which is again of the $\dfrac{0}{0}$ type.

$\therefore \dfrac{(1 - \sec^2 x)'}{(1 - \cos x)'} = - \dfrac{2\,(\sec x)\,(\sec x)\,(\tan x)}{\sin x}$, by chain rule

$$= - \dfrac{2 \dfrac{1}{\cos x} \cdot \dfrac{1}{\cos x} \cdot \dfrac{\sin x}{\cos x}}{\sin x}$$

$$= - \dfrac{2}{\cos^3 x} .$$

Moveover, $\lim\limits_{x \to 0} - \dfrac{2}{\cos^3 x} = -2$, since $\lim\limits_{x \to 0} \cos x = 1$.

Hence, by L'Hôpital's Rule

$$\lim\limits_{x \to 0} \dfrac{x - \tan x}{x - \sin x} = \lim\limits_{x \to 0} \dfrac{1 - \sec^2 x}{1 - \cos x}$$

$$= \lim\limits_{x \to 0} - \dfrac{2}{\cos^3 x}$$

$$= -2$$

19. (C)

$$\lim\limits_{n \to \infty} \dfrac{6n^2}{200 - 4n + kn^2} = \lim\limits_{n \to \infty} \dfrac{\dfrac{6n^2}{n^2}}{\dfrac{200}{n^2} - \dfrac{4}{n} + k} ,$$

by dividing both the numerator and the denominator by n^2.

Therefore,

$$\lim_{n \to \infty} \frac{6n^2}{200 - 4n + kn^2} = \lim_{n \to \infty} \frac{6}{\dfrac{200}{n^2} - \dfrac{4}{n} + k}$$

$$= \frac{6}{k}, \text{ since } \frac{200}{n^2} \to 0 \text{ and } \frac{4}{n} \to 0.$$

From what is given,
$$\frac{6}{k} = \frac{1}{2}$$

$$\Rightarrow k = 12$$

20. (B)

$$\lim_{n \to \infty} \left(1 + \frac{1}{n}\right)^{kn} = \lim_{n \to \infty} \left[\left(1 + \frac{1}{n}\right)^n\right]^k$$

$$= \left[\lim_{n \to \infty} \left(1 + \frac{1}{n}\right)^n\right]^k, \qquad \begin{array}{l}\text{by property of} \\ \text{limit of powers}\end{array}$$

$$= e^k, \text{ since } \lim_{n \to \infty} \left(1 + \frac{1}{n}\right)^n = e.$$

But we are given that

$$\lim_{n \to \infty} \left(1 + \frac{1}{n}\right)^{kn} = \frac{1}{e^2}$$

$$= e^{-2}$$

$$\Rightarrow e^k = e^{-2}$$

$$\Rightarrow k = -2$$

188

21.   (D)
$$y = - \frac{1}{\sqrt{x^2 + 1}}$$

$$= - (x^2 + 1)^{-1/2}$$

$$\Rightarrow \frac{dy}{dx} = (-1)\left(-\frac{1}{2}\right)(x^2 + 1)^{-3/2}(2x), \text{ by the chain rule,}$$

$$= \frac{x}{(x^2 + 1)^{3/2}}$$

22.   (E)
$$\frac{dy}{dx} = \frac{1}{(x + 1)(x + 2)}\left[\frac{d}{dx}((x + 1)(x + 2))\right],$$
by the chain rule

$$= \frac{1}{(x + 1)(x + 2)}((x + 2) + (x + 1)),$$
by the product rule

$$= \frac{1}{x + 1} + \frac{1}{x + 2}.$$

Alternatively, we can use the fact that $\ln(ab) = \ln(a) + \ln(b)$, so that $y = \ln[(x + 1)(x + 2)]$

$$= \ln(x + 1) + \ln(x + 2)$$

$$\Rightarrow \frac{dy}{dx} = \frac{1}{x + 1} + \frac{1}{x + 2}.$$

23. **(B)**

$$f'(x) = kae^{kx}$$

$$\Rightarrow \frac{f'(x)}{f(x)} = \frac{kae^{kx}}{ae^{kx}}$$

$$= k$$

$$\Rightarrow k = -\frac{5}{2}$$

24. **(B)**

$$\frac{\text{height}}{s} = \sin 60°$$

$$\Leftrightarrow \text{height} = s(\sin 60°)$$

$$= s\frac{\sqrt{3}}{3}$$

$$\Rightarrow \text{Area} = \frac{1}{2}h(S)$$

$$= \left(\frac{1}{2}\right)\frac{\sqrt{3}(s)(s)}{2}, \quad \text{substituting in for } h.$$

$$= \frac{\sqrt{3}\,s^2}{4}$$

$$\Rightarrow s\,\frac{dA}{ds} = \frac{\sqrt{3}\,S}{2}$$

Therefore, when $s = 2$, $\dfrac{dA}{ds} = \dfrac{2\sqrt{3}}{2}$

$$= \sqrt{3}$$

$$\approx 1.73$$

25.   (D)
$$f(x) = e^{\frac{x^3}{3} - x}$$

$$\Rightarrow f'(x) = (x^2 - 1)\, e^{\frac{x^3}{3} - x}, \qquad \text{by the chain rule.}$$

We see $e^{\frac{x^3}{3} - x} > 0$ for every real number x

and $x^2 - 1 = (x + 1)(x - 1)$

$$\Rightarrow x^2 - 1 < 0 \quad \text{when } x \in (-1, 1) \text{ and}$$
$$x^2 - 1 > 0 \quad \text{when } |x| > 1.$$

Hence,
$$f'(x) = (x^2 - 1)\, e^{\frac{x^3}{3} - x} < 0 \quad \text{for } x \in (-1, 1) \text{ and}$$

$$f'(x) = (x^2 - 1)\, e^{\frac{x^3}{3} - x} > 0 \quad \text{for } |x| > 1.$$

(D) is the answer.

26.   (C)
$$f(x) = x - \frac{1}{x}$$

$$\Rightarrow f\left(\frac{1}{x}\right) = \frac{1}{x} - \frac{1}{\frac{1}{x}}$$

$$= \frac{1}{x} - x$$

$$= -\left(x - \frac{1}{x}\right), \qquad \text{factoring out } -1.$$

$$= -f(x), \qquad \text{so (A) is true.}$$

$$f\left(\frac{1}{x}\right) = -f(x) = -x + \frac{1}{x}$$

$$= -x - \frac{1}{-x} = f(-x), \qquad \text{so (B) is true.}$$

$$f(-x) = (-x) - \frac{1}{(-x)}$$

$$= -x + \frac{1}{x}$$

$$= -\left(x - \frac{1}{x}\right)$$

$$= -f(x), \qquad\qquad \text{so (C) is false.}$$

27. (B)

$$x + y = 20$$

$$\Rightarrow y = 20 - x$$

$$\Rightarrow xy = x(20 - x)$$

$$= 20x - x^2$$

Now let $f(x) = 20x - x^2$

$$\Rightarrow f'(x) = 20 - 2x$$

$$\Rightarrow f'(x) = -2(x - 10)$$

$$f'(x) = 0 \Rightarrow 0 = -2(x - 10)$$

$$\Rightarrow x = 10$$

We see that $f''(x) = -2 < 0$, so we have a maximum at the critical point $x = 10$. Since $x + y = 20$, we see $y = 10$.

28.   **(B)**

$$f(x) = 2\sqrt{x}$$

$$= 2x^{1/2}$$

$$\Rightarrow f'(x) = (2)\left(\frac{1}{2}\right)x^{\frac{1}{2}-1}$$

$$= \frac{1}{\sqrt{x}}$$

We are given that $f'(x_0) = 2f'(4)$

$$\Rightarrow \frac{1}{\sqrt{x_0}} = 2\frac{1}{\sqrt{4}}$$

$$\Rightarrow \frac{1}{\sqrt{x_0}} = 1$$

$$\Rightarrow x_0 = 1$$

29.   **(A)**

Draw the graphs of $f(x)$ and $g(x)$.

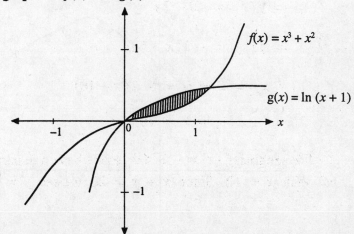

Reset viewing window to $[0, 1] \cdot [0, 1]$ and trace to the intersection point of $f(x)$ and $g(x)$, which should turn out to be $x = 0.5238, y = 0.418$.

The area enclosed is the integral $\displaystyle\int_{0}^{0.5238} \ln(x + 1) - (x^3 + x^2)$,

using your calculator, which can easily be computed. The number is 0.0513.

193

30. (B)
Draw the graphs of $y = \sin 3x \cos x$.

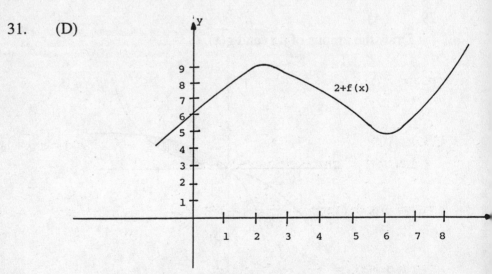

Tracing the first value of x when $y = 0$ gives the x-intercept 1.03.
Then the area is the integral

$$\int_0^{1.03} \sin 3x \quad \cos x$$

or *fnInt* $(\sin 3x \cos x, x, 0, 1.03)$ which equals 0.56.

31. (D)

As you can see from the above graph, lifting the graph vertically upwards or downwards does not change the nature of the concavity of the inflection points.

194

32. (D)

$$f(x) = g(x) - \frac{1}{g(x)}$$

$$\Rightarrow f'(x) = \frac{d}{dx}(f(x)) = g'(x) - \frac{d}{dx}\left(\frac{1}{g(x)}\right)$$

$$= g'(x) - \left(\frac{0 \cdot g(x) - g'(x) \cdot 1}{[g(x)]^2}\right),$$

by the quotient rule.

$$= g'(x) - \left(-\frac{g'(x)}{[g(x)]^2}\right)$$

$$= g'(x) + \frac{g'(x)}{[g(x)]^2}$$

As a result, $f'(0) = g'(0) + \dfrac{g'(0)}{[g(0)]^2}$

$$= 2 + \frac{2}{9} = \frac{20}{9} \approx 2.222$$

Calculator: $20 + 9 = \approx 2.222$

33. (C)

Let $s(t) = $ distance traveled in time t.

$$\Rightarrow \text{acceleration} = \frac{d^2 s}{dt^2} = (4t + 1)^{1/2}$$

and $v(t) = \dfrac{ds}{dt} = \displaystyle\int a(t)\,dt$

$$= \int (4t + 1)^{1/2}\,dt$$

$$\Rightarrow s(t) = \frac{1}{60}(4t+1)^{5/2} - \frac{9}{2}t + C$$

$$\Rightarrow s(2) - S(0) = \left(\frac{1}{60}(9)^{5/2} - 9 + C\right) - \left(\frac{1}{60} + C\right)$$

$$= \frac{3^5}{60} - 9 - \frac{1}{60}$$

$$= \frac{243}{60} - 9 - \frac{1}{60}$$

$$= 4 + \frac{2}{60} - 9$$

$$= -\frac{149}{30}$$

34.   (A)

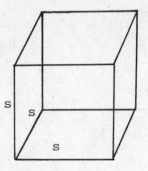

Let the edge of the cube at time t be $S(t)$.

$$\Rightarrow V = S^3$$

$$\Rightarrow \frac{dV}{dt} = 3[S]^2 \frac{dS}{dt} \quad \text{by the chain rule.}$$

But we are given that $\frac{dV}{dt}$ 300 in³/ min.

So 300 in³/ min. $= 3S^2 \frac{dS}{dt}$

$$\Rightarrow \frac{dS}{dt} = \frac{300}{3S^2} \text{ in}^3/\text{min.}$$

Therefore, when the edge is 20 inches long,

$$\frac{dS}{dt} = \frac{300 \ \text{in}^3/\text{min.}}{(3)(20)(20)\text{in}^2}$$

$$= \frac{1}{4} \ \text{in}/\text{min.}$$

35.   (E)

Using your calculator,

$$fnInt\left(\frac{6(x \wedge 3 + 1)}{x + 1}, x, -1, 1\right),$$

which gives 16.

36.   (C)

$$f(x) = x^3 - 2x$$

$$\Rightarrow f'(x) = 3x^2 - 2$$

Therefore, $f'(x) = 0$

$$\Rightarrow 3x^2 - 2 = 0$$

$$\Rightarrow x = \pm \sqrt{\frac{2}{3}}$$

Also $f''(x) = 6x$

$$\Rightarrow f''\left(\sqrt{\frac{2}{3}}\right) = 6\sqrt{\frac{2}{3}} > 0$$

and $f''\left(-\sqrt{\frac{2}{3}}\right) = -6\sqrt{\frac{2}{3}} < 0.$

Hence, f has a relative Max. at $x = -\sqrt{\dfrac{2}{3}}$ and a relative Min. at $x = \sqrt{\dfrac{2}{3}}$ by the second derivative test.

37.   (A)

You can directly solve this problem by using the calculator. For example,

$$\text{der } 1 \left(256\, x^\wedge (-0.5) + 64\, x^\wedge 0.5 + 3 x^\wedge \left(\frac{2}{3}\right), x, 64 \right)$$

should get 4.25.

38.   (B)

Draw the graph of $y = x^2 (1 - x^{-1})$.

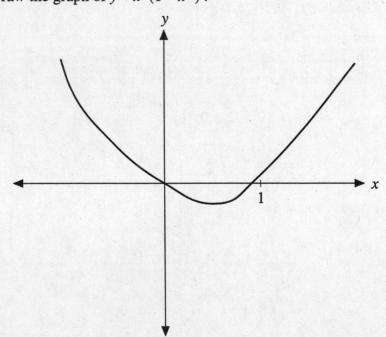

Set the viewing window to $[0, 1] \cdot [-1, 1]$. By tracing y to the lowest value, you can get $y = -0.25$.

39.   (B)

$$\int_a^b f(x)\ dx = \int_a^b f(x)\ dx + \int_b^c f(x)\ dx + \int_c^d f(x)\ dx .$$

Since $f(x)$ is non-negative in $[a, b]$ and $[c, d]$ the integral of $f(x)$ in $[a, b]$ and $[c, d]$ corresponds with the areas of regions A and C respectively.

Since $f(x)$ is non-positive in $[b, c]$, the integral is actually the negative of the area of region B.

Hence, $\displaystyle\int_a^d f(x)\ dx = 3 - \left(1\frac{1}{2}\right) + 2 = 3\frac{1}{2} = \frac{7}{2}$

40.   (C)

Use integration by parts:

Let $u = x$ and $dv = \cos (4x)\ dx$

$\Rightarrow du = dx$ and $V = \dfrac{1}{4} \sin (4x)$

$\Rightarrow \displaystyle\int x\cos (4x)\ dx = \frac{x}{4} \sin (4x) - \int \frac{1}{4} \sin (4x)\ dx$

$\qquad\qquad = \dfrac{x}{4} \sin (4x) + \dfrac{1}{16} \cos (4x) + c$

ADVANCED PLACEMENT CALCULUS AB EXAM IV

DETAILED EXPLANATIONS OF ANSWERS

1. (A)

$$\lim_{x \to \frac{\pi}{4}} \left(\frac{1}{2} + \frac{1}{2} \cos (2x) \right)$$

$$= \lim_{x \to \frac{\pi}{4}} \frac{1}{2} + \lim_{x \to \frac{\pi}{4}} \frac{1}{2} \cos (2x) \text{ , by addition rule.}$$

$$= \frac{1}{2} + \frac{1}{2} \lim_{x \to \frac{\pi}{4}} \cos (2x) \text{ , since } \lim_{x \to a} c \, f(x) = c \lim_{x \to a} f(x)$$

$$= \frac{1}{2} + \frac{1}{2} \cdot \cos \left(2 \cdot \frac{\pi}{4} \right) \text{ , since } \cos x \text{ is a continuous function.}$$

$$= \frac{1}{2} + \frac{1}{2} \cos \frac{\pi}{2}$$

$$= \frac{1}{2} + \frac{1}{2} \cdot 0 \text{ , since } \cos \frac{\pi}{2} = 0,$$

$$= \frac{1}{2}$$

(B)

Average value $= \dfrac{1}{\pi - 0} \displaystyle\int_0^\pi \left(\dfrac{1}{2} + \dfrac{1}{2} \cos (2x) \right) dx$

$= \dfrac{1}{\pi} \left(\displaystyle\int_0^\pi \dfrac{1}{2} dx + \dfrac{1}{2} \displaystyle\int_0^\pi \cos (2x) \, dx \right)$

$= \dfrac{1}{\pi} \left(\dfrac{1}{2} \cdot \pi + \dfrac{1}{2} \cdot \dfrac{1}{2} \sin (2x) \Big|_0^\pi \right)$

$= \dfrac{1}{\pi} \left(\dfrac{\pi}{2} + \dfrac{1}{4} (\sin 2\pi - \sin 0) \right)$

$= \dfrac{1}{2}$, since $\sin 2\pi = \sin 0 = 0$

(C)

First, $f(b) = \dfrac{1}{2} + \dfrac{1}{2} \cos (2b)$

and $f(a) = \dfrac{1}{2} + \dfrac{1}{2} \cos (2a)$.

$\Rightarrow f(b) - f(a) = \dfrac{1}{2} (\cos (2b) - \cos (2a))$.

By the Mean Value Theorem,

$f(b) - f(a) = (b - a) \, f'(c)$ for some $a < c < b$.

But, $f'(c) = -\sin (2c)$, since $f'(x) = \dfrac{1}{2} \cdot 2 (-\sin 2x)$, by the chain rule.

$\Rightarrow | f(b) - f(a) | = |-\sin (2c)| \, |b - a|$

$\Rightarrow | f(b) - f(a) | \le |b - a|$, since $|-\sin (2c)| \le 1$.

2.   (A)

Use a viewing window of $[-5, 5] \cdot [-5, 40]$ and draw the graphs of $f(x)$ and $g(x)$.

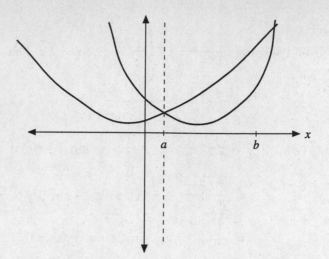

By tracing on the graphs to the integrational points, you can find points $a = 0.2$ and $b = 4.81$

(B)

Draw the graphs of $f'(x)$ and $g'(x)$.

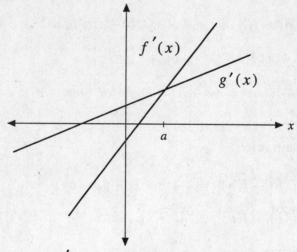

By tracing to where $f'(x) = g'(x)$, you can find $a = 2.57$.

(C)

Finding $\displaystyle\int_{0.2}^{4.81} ((x^2 + 2x + 4) - (2x^2 - 2x + 5)) \ dx$

by using the calculator.

fnInt $(-x^{\wedge}2 + 4x - 1, x, 0.2, 4.81)$

results in 4.49 .

202

3.   (A) (i)

f is an even function

$\Rightarrow f(x) \quad = f(-x)$ for every x

$\Rightarrow f'(x) \quad = -f'(-x)$, by chain rule

$\Rightarrow f'(2) \quad = -f'(-2)$

$\Rightarrow f'(-2) = -f'(2)$

$\Rightarrow f'(-2) = -5$, since $f'(2) = 5$.

       (ii)
$f'(x) = -f'(-x)$

$\Rightarrow f'(0) = -f'(-0)$

$\Rightarrow f'(0) = -f'(0)$

$\Rightarrow f'(0) + f'(0) = 0$, adding $f'(0)$ to both sides.

$\Rightarrow 2 f'(0) = 0$

$\Rightarrow f'(0) = 0$, multiplying both sides by $\dfrac{1}{2}$.

     (B)
Slope of $L_1 = f'(2) = 5$

Slope of $L_2 = f'(-2) = -5$

Moreover,
L_1 passes through $(2, 1)$ and
L_2 passes through $(-2, 1)$, since $f(2) = f(-2) = 1$, because f is
an even function.

Therefore, the point slope form of the equation of
L_1 is $y - 1 = 5(x - 2)$ and that of
L_2 is $y - 1 = -5(x + 2)$.

$\Rightarrow y = 5(x - 2) + 1$

and $y = -5(x + 2) + 1$

Solving these 2 equations simultaneously, we have:

$5(x - 2) + 1 = -5(x + 2) + 1$

$\Rightarrow 5(x - 2) + 5(x + 2) = 0$

$\Rightarrow 10x = 0$

$\Rightarrow x = 0$

$\Rightarrow y = 5(x - 2) + 1$

$\qquad = 5(0 - 2) + 1$

$\qquad = 5(-2) + 1$

$\qquad = -10 + 1$

$\Rightarrow y = -9$

So $P = (0, -9)$

4.   (A)

h is a constant function and the point $(2, -2)$ lies on the graph of h. So h is the constant function $h(x) = -2$.

By similar reasoning $g(x) = -5$.

$$\Rightarrow h(x) - g(x) = -2 - (-5)$$

$$= 3 \text{ , a positive constant.}$$

Therefore, $(h - g) f = 3f$.

Since $\frac{d}{dx}(3f) = 3f'$ and $\frac{d^2}{dx^2} f = 3f''$, then

$3f$, like f, is concave upward in the intervals

$(-\frac{1}{4}, 3) \cup (6, 9)$;

and $3f$ is concave downward in the intervals

$(-5, -\frac{1}{4}) \cup (3, 6)$.

(B)

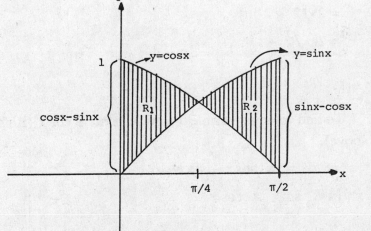

$(h-g) f=3f$

5.   (A)

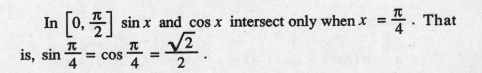

In $\left[0, \dfrac{\pi}{2}\right]$ $\sin x$ and $\cos x$ intersect only when $x = \dfrac{\pi}{4}$. That is, $\sin \dfrac{\pi}{4} = \cos \dfrac{\pi}{4} = \dfrac{\sqrt{2}}{2}$.

205

(B)

In the inverval $\left[0, \dfrac{\pi}{4}\right]$, $\cos x \geq \sin x$.

In the interval $\left[\dfrac{\pi}{4}, \dfrac{\pi}{2}\right]$, $\sin x \geq \cos x$. (Refer to the figure above).

Therefore, Area =

$$\int_0^{\frac{\pi}{4}} (\cos x - \sin x)\, dx + \int_{\frac{\pi}{4}}^{\frac{\pi}{2}} (\sin x - \cos x)\, dx$$

$$= (\sin x + \cos x)\Big|_0^{\frac{\pi}{4}} + (-\cos x - \sin x)\Big|_{\frac{\pi}{4}}^{\frac{\pi}{2}}$$

$$= \left((\sin \tfrac{\pi}{4} + \cos \tfrac{\pi}{4}) - (\sin 0 + \cos 0) \right) +$$

$$\left((-\cos \tfrac{\pi}{2} - \sin \tfrac{\pi}{2}) - (-\cos \tfrac{\pi}{4} - \sin \tfrac{\pi}{4}) \right)$$

$$= \frac{\sqrt{2}}{2} + \frac{\sqrt{2}}{2} - (1) + \left(-1 - \left(-\frac{\sqrt{2}}{2} - \frac{\sqrt{2}}{2} \right) \right),$$

since $\sin 0 = \cos \dfrac{\pi}{2} = 0$.

$$= \sqrt{2} - 1 - 1 + \sqrt{2}$$

$$= 2\sqrt{2} - 2$$

$$= 2(\sqrt{2} - 1).$$

(C)

To find volume, we note that in R_1 $\sin x \leq \cos x$ (Refer to figure above).

$$\Rightarrow dV = \pi (\cos x - \sin x)^2\, dx\ ;$$

In R_2 $\sin x \geq \cos x$

$$\Rightarrow dV = \pi (\sin x - \cos x)^2\, dx$$

Therefore, volume =

$$= \int_0^{\frac{\pi}{4}} \pi (\cos x - \sin x)^2\, dx + \int_{\frac{\pi}{4}}^{\frac{\pi}{2}} \pi (\sin x - \cos x)^2\, dx$$

Equivalently, volume $= \pi \int_0^{\frac{\pi}{2}} (\cos x - \sin x)^2 \, dx$.

Since $(\cos x - \sin x)^2 = (\sin x - \cos x)^2$, we can combine the 2 intergrals into one.

$$= \pi \int_0^{\frac{\pi}{2}} (\cos^2 x - 2 \sin x \cos x + \sin^2 x) \, dx$$

$$= \pi \int_0^{\frac{\pi}{2}} (1 - \sin(2x)) \, dx, \quad \text{using the identities}$$

$\sin^2 x + \cos^2 x = 1$ and $2 \sin x \cos x = \sin 2x$.

6.   (A)
$f'(x) = 3x^2 + 2ax + b$.

Since f has a relative minimum at $x = 4$, then

$f'(4) = 3(4^2) + 2a(4) + b = 0$

$\Rightarrow 48 + 8a + b = 0 \qquad\qquad (1)$

Since f has an inflection point at $x = 1$

$\Rightarrow f''(1) = 0$ but $f''(x) = 6x + 2a$

$f''(1) = 0 \Rightarrow 6 + 2a = 0$

$\qquad\qquad \Rightarrow a = -3 \qquad\qquad (2)$

We substitute Eq. (2) into Eq. (1) to find b:

$48 + 8(-3) + b = 0$

$24 + b = 0$

$b = -24$.

     (B)
When we plug $a = -3$ and $b = -24$ into
$f(x) = x^3 + ax^2 + bx + c$ we get

$f(x) = x^3 - 3x^2 - 24x + c$

207

But $f(0) = 1$

$\Rightarrow 1 = f(0) = 0^3 - 3(0)^2 - 24(0) + c = c$

Hence $f(x) = x^3 = 3x^2 - 24x + 1$

$\Rightarrow f'(x) = 3x^2 - 6x - 24$

So $f'(x) = 0$

$\Rightarrow 3x^2 - 6x - 24 = 0$

$\Rightarrow x^2 - 2x - 8 = 0$ (multiplying both sides by $\frac{1}{3}$)

$\Rightarrow (x - 4)(x + 2) = 0$

$\Rightarrow x = 4$ or $x = -2$

Moreover, $f''(x) = 6x - 6$

Plugging in the critical points $x = 4$, $x = -2$, we see that

$\Rightarrow f''(4) = 18 > 0$ and $f''(-2) = -18 < 0$

Hence f has a relative Max. at $x = -2$
and a relative Min. at $x = 4$.

So the relative Maximum of f is

$f(-2) = -8 - 12 + 48 + 1$

$\qquad = 29$.

THE ADVANCED PLACEMENT EXAMINATION IN

CALCULUS AB

TEST V

ADVANCED PLACEMENT CALCULUS AB EXAM V

SECTION I

PART A

Time: 45 minutes
25 questions

DIRECTIONS: Each of the following problems is followed by five choices. Solve each problem, select the best choice, and blacken the correct space on your answer sheet. Calculators may not be used for this section of the exam.

NOTE:
Unless otherwise specified, the domain of function f is assumed to be the set of all real numbers x for which $f(x)$ is a real number.

1. If $3x^2 - x^2y^3 + 4y = 12$ determines a differentiable function such that $y = f(x)$, then $\dfrac{dy}{dx} =$

    (A) $\dfrac{-3x + 2xy^3}{2}$

    (B) $\dfrac{-6x + 2xy^3}{-3x^2y^2 + 4}$

    (C) $\dfrac{-3x + 2xy^3 + 6}{2}$

    (D) $\dfrac{-6x + 2xy^3 + 3x^2y^2}{4}$

    (E) $\dfrac{-6x + 2xy^3}{3x^2y^2 + 4}$

211

2. $\displaystyle\int_1^4 \frac{5x^2 - x}{2\sqrt{x}}\, dx =$

(A) 29

(D) $\dfrac{311}{12}$

(B) $\dfrac{113}{6}$

(E) $\dfrac{100}{3}$

(C) $\dfrac{86}{3}$

3. The area in the first quadrant that is enclosed by the graphs of $x = y^3$ and $x = 4y$ is

(A) 4

(D) 1

(B) 8

(E) 0

(C) -4

4. For what value of c is $f(x) = \begin{cases} 3x^2 + 2, & x \geq -1 \\ -cx + 5, & x < -1 \end{cases}$ continuous?

(A) 3

(D) None

(B) -3

(E) 0

(C) 6

212

5. $\lim\limits_{x \to \infty} \sqrt[3]{\dfrac{8 + x^2}{x(x+1)}} =$

(A) 0

(D) 1

(B) 2

(E) Does not exist

(C) $\sqrt[3]{9}$

6. If $y = \tan(\operatorname{arcsec} x)$, then $\dfrac{dy}{dx} =$

(A) $\sqrt{x^2 - 1}$

(D) $\dfrac{x}{\sqrt{x^2 - 1}}$

(B) $\dfrac{x}{\sqrt{1 + x^2}}$

(E) $\dfrac{x}{\sqrt{1 - x^2}}$

(C) $\dfrac{\sqrt{x^2 - 1}}{x}$

7. Let $f(x) = 2x^5 - x^3 + x^2 + 2$ and $g(x) = f^{-1}(x)$.
If $f(1) = 4$ then $g'(4) =$

(A) 9

(D) $\dfrac{1}{2002}$

(B) $\dfrac{1}{9}$

(E) $\dfrac{1}{2376}$

(C) $\dfrac{1}{4}$

8. $\displaystyle\lim_{x\to 0^-} (1-x)^{2/x} =$

(A) e^{-2}

(D) e^2

(B) -2

(E) Does not exist

(C) 2

9. If $f(x) = |x|$ for all real numbers x, then $f'(x)$ is a real number for

(A) $x < 0$ only

(D) $x \neq 0$ only

(B) $x > 0$ only

(E) All real numbers x

(C) $x = 0$ only

10. Note that $\sin \dfrac{\pi}{2} \neq \sin \left(\dfrac{\pi}{2}\right)^{\circ}$ since $\sin \dfrac{\pi}{2} = 1$ and $\sin \left(\dfrac{\pi}{2}\right)^{\circ} = \sin (1.578^{\circ}) \cong .03$. $\dfrac{d}{dx}(\sin(x^{\circ})) =$

(A) $\cos (x^{\circ})$

(D) $\dfrac{\pi}{180} \cos (x^{\circ})$

(B) $\dfrac{\pi}{2} \cos (x^{\circ})$

(E) $\dfrac{\pi}{180} \cos x$

(C) $\cos \left(\dfrac{\pi x}{180}\right)$

214

11. $\displaystyle\int \frac{\log(x^3 \cdot 10^x)}{x}\, dx =$ (<u>Note</u>: log stands for $\log_{10}$ and ln stands for $\log_e$.)

(A) $\displaystyle\frac{3\ln 10}{2}(\log x)^2 + x + C$

(B) $\displaystyle\frac{3}{2\ln 10}(\ln x)^2 + x + C$

(C) $\displaystyle\frac{6\log x}{\ln 10} + x + C$

(D) $\displaystyle\frac{3(\log x)^2}{\ln 10} + x + C$

(E) $3\ln 10\,(\ln x^2) + x + C$

12. $f(x) = 2x^3 - 9x^2 + 12x - 3$ is decreasing for

(A) $x < 2$

(B) all values of x

(C) $x < 1$ and $x > 2$

(D) $1 < x$

(E) $1 < x < 2$

215

13. The equation of each horizontal asymptote of
$$f(x) = \frac{|x|}{|x| + x}$$

(A) $y = \frac{1}{2}$

(D) $y = -1$

(B) $y = 0$

(E) $x = 0$

(C) $y = 1$

14. Find the point on the parabola $y = x^2$ which is closest to the point $(6, 3)$.

(A) $(1\frac{1}{2}, 2\frac{1}{4})$

(D) $(3, 9)$

(B) $(2\frac{1}{2}, 6\frac{1}{4})$

(E) $(1\frac{3}{4}, 3\frac{1}{16})$

(C) $(2, 4)$

15. $\displaystyle\lim_{x \to -1} \frac{\sqrt{x^2 + 3} - 2}{x + 1} =$

(A) 0

(D) 2

(B) -2

(E) Does not exist

(C) $-\frac{1}{2}$

16. The area bounded by the parabola $y^2 = 2x - 2$ and the line $y = x - 5$ is

(A) 22

(D) $\dfrac{52}{3}$

(B) 18

(E) $\dfrac{26}{3}$

(C) $\dfrac{14}{3}$

17. Population growth in a certain bacteria colony is best described by the equation $y = t^2 e^{3t^2 + \sqrt{t}}$, where t is in hours. The rate of growth of the colony at $t = 1$ is

(A) 464.084

(D) 300.290

(B) 245.692

(E) 545.982

(C) 409.486

18. $\displaystyle \int \dfrac{e^{2x}}{e^x - 3} \, dx =$

(A) $(e^x + 3) + 3 \ln|e^x + 3| + C$

(B) $e^x - 3 \ln|e^x - 3| + C$

(C) $(e^x - 3) + 3 \ln|e^x - 3| + C$

(D) $(e^x - 3) - \dfrac{3}{(e^x - 3)^2} + C$

(E) $(e^x - 3) + \dfrac{3}{(e^x - 3)^2} + C$

19. If $xy - x = 2y - 5$, then which of the following must be true?

   I. The relation is a function of x.

   II. The domain is $\{ x \mid x \neq 2 \}$.

   III. The range is all reals.

   (A) I only

   (B) II only

   (C) I and II only

   (D) I and III only

   (E) I, II and III

20. If $h(x) = \sqrt{x^2 - 4}$, $x \leq -2$, then $h^{-1}(x) =$

   (A) $\sqrt{x^2 + 4}$

   (B) $-\sqrt{x^2 + 4}$

   (C) $-\sqrt{x + 4}$

   (D) $\sqrt{x + 4}$

   (E) $-\sqrt{x^2 - 4}$

21. Which of the following is the graph of $y = 1 + 2^{x+3}$?

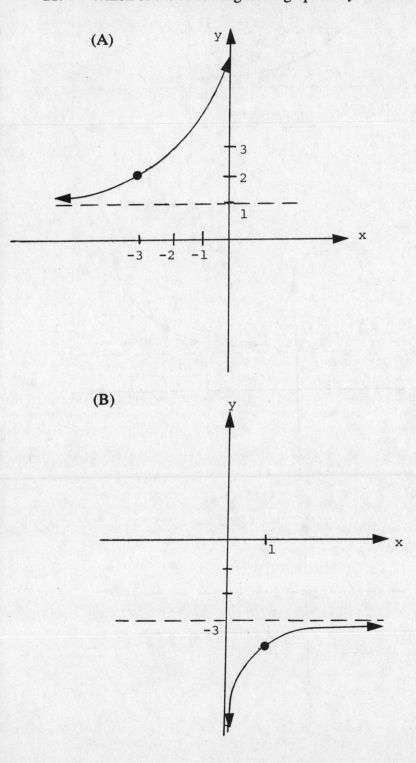

(A)

(B)

(C)

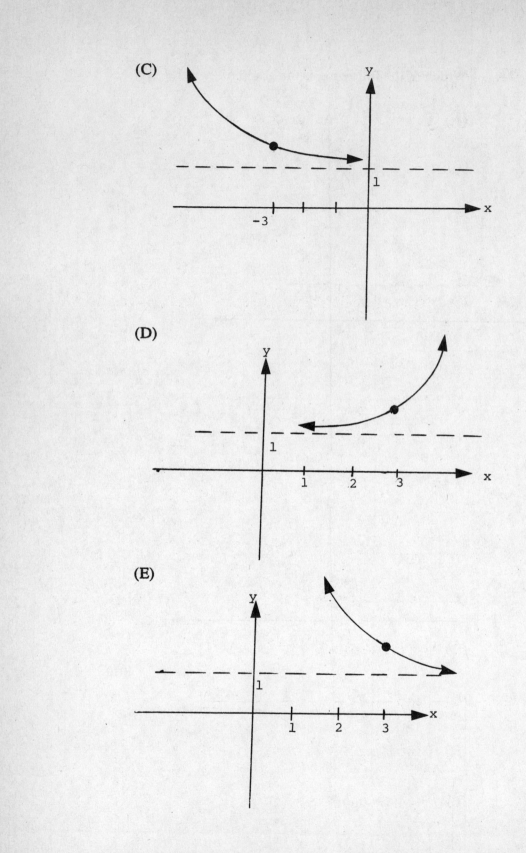

(D)

(E)

22. $\lim\limits_{x \to 0} \dfrac{\sin 2x}{x \cos x} =$

(A) 0

(D) 2

(B) 1

(E) Does not exist

(C) $\dfrac{1}{2}$

23. Find the area bounded by the curve $xy = 4$, the x-axis, $x = e$ and $x = 2e$.

(A) 4.000

(D) 1.386

(B) 0.693

(E) 2.000

(C) 2.773

24. $\dfrac{\cos 2\beta - \sin 2\beta}{\sin \beta \cos \beta} =$

(A) $\cot \beta - \tan \beta - 2$

(B) $\cot \beta + \tan \beta + 2$

(C) $\tan \beta - \cot \beta - 2$

(D) $\tan \beta + \cot \beta - 2$

(E) $- \cot \beta - \tan \beta - 2$

25. The figure shown is the graph of a function f. Identify the equation which corresponds to this graph of f for x such that

$$-\frac{\pi}{3} \le x \le \frac{11\pi}{3} .$$

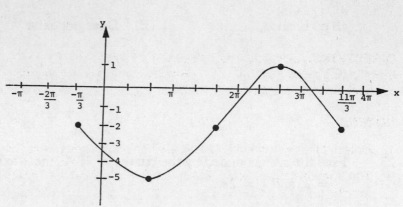

(A) $f(x) = -3\sin 2\left(x - \frac{\pi}{3}\right) - 2$

(B) $f(x) = -3\sin\frac{1}{2}\left(x - \frac{\pi}{3}\right) + 2$

(C) $f(x) = 3\sin\frac{1}{2}\left(x + \frac{\pi}{3}\right) - 2$

(D) $f(x) = -3\sin\frac{1}{2}\left(x + \frac{\pi}{3}\right) - 2$

(E) $f(x) = 3\sin\frac{1}{2}\left(x - \frac{\pi}{3}\right) - 2$

PART B

Time: 45 minutes
15 questions

DIRECTIONS: Calculators may be used for this section of the test. Each of the following problems is followed by five choices. Solve each problem, select the best choice, and blacken the correct space on your answer sheet.

NOTES:

1. Unless otherwise specified, answers can be given in unsimplified form.

2. The domain of function f is assumed to be the set of all real numbers x for which $f(x)$ is a real number.

26. If $y = \dfrac{2(x-1)^2}{x^2}$, then which of the following must be true?

   I. The range is $\{\, y \mid y \geq 0 \,\}$.

   II. The y-intercept is 1.

   III. The horizontal asymptote is $y = 2$.

   (A) I only   (D) I and II only

   (B) II only   (E) I and III only

   (C) III only

27. If $g(x + 3) = x^2 + 2$, then $g(x)$ equals

   (A) $\sqrt{x - 3}$ (D) $\sqrt{x - 3} - 2$

   (B) $(x - 3)^2 + 2$ (E) $\sqrt{x - 3} + 2$

   (C) $(x - 3)^2 - 2$

223

28. The slope of the tangent line to the graph $y = \dfrac{x^2}{\sqrt[3]{3x^2 + 1}}$ at $x = 1$ is approximately

(A) 0.945   (D) 1.191

(B) 2.381   (E) 1.575

(C) 1.890

29. If x and y are functions of t which satisfy the relation $x^4 + xy + y^4 = 1$, then $\dfrac{dy}{dt} =$

(A) $\left(\dfrac{4x + y}{x + 4y^3}\right)\dfrac{dx}{dt}$

(B) $-\left(\dfrac{4x^3 + y}{x + 4y^3}\right)\dfrac{dx}{dt}$

(C) $-\left(\dfrac{x + 4y^3}{4x^3 + y}\right)\dfrac{dx}{dt}$

(D) $-\left(\dfrac{4x^3 + y}{x + 4y}\right)\dfrac{dx}{dt}$

(E) $\left(-\dfrac{4x^3 + y}{x + 4y^3}\right)\dfrac{dx}{dt}$

30. If $g(u) = \sqrt{u^3 + 2}$, $f(1) = 2$ and $f'(1) = -5$, then $\dfrac{d}{dx}\left(g(f(x))\right)$ at $x = 1$ is approximately

(A) 9.487   (D) –9.487

(B) 18.974   (E) 4.330

(C) –4.330

224

31. If $y = \dfrac{1}{\sin(t + \sqrt{t})}$, then $y'(1)$ is approximately:

(A) –0.002

(D) 410.267

(B) –1,641.070

(E) –0.070

(C) –1,230.802

32. Find $\displaystyle\int \arctan x \, dx$ using integration by parts.

(A) $\arctan x + \ln(1 + x^2) + C$

(B) $x \arctan x + C$

(C) $\dfrac{1}{(1 + x^2)} + C$

(D) $x \arctan x - \dfrac{1}{2} \ln(1 + x^2) + C$

(E) $\ln(1 + x^2) + x \arctan x + C$

33. If $y = -\ln\left| \dfrac{1 + \sqrt{1 - x^2}}{x} \right|$, then $\dfrac{dy}{dx} =$

(A) $\dfrac{1}{x\sqrt{1 - x^2}}$

(D) $\dfrac{1}{\sqrt{1 - x^2}}$

(B) $\dfrac{1}{x} - \dfrac{1}{\sqrt{1 - x^2}}$

(E) $\dfrac{x}{\sqrt{1 - x^2}}$

(C) $\dfrac{x + 1}{x\sqrt{1 - x^2}}$

34. Let $F = \dfrac{6000\,k}{k\sin\theta + \cos\theta}$ where k is a constant. For which value of θ is $\dfrac{dF}{d\theta} = 0$ for $\dfrac{-\pi}{2} < \theta < \dfrac{\pi}{2}$?

(A) $\theta = \arctan\dfrac{1}{k}$

(D) $\theta = \arctan k$

(B) $\theta = 0$

(E) $\theta = \operatorname{arccot} k$

(C) $\theta = \dfrac{\pi}{2}$

35. If $f(x) = \dfrac{1}{\sin(x + \sqrt{x})}$, calculate $\displaystyle\int_0^1 f(x)$.

(A) 2.57

(D) 98.1

(B) 10.26

(E) 1.65

(C) –3.1

36. $f(x) = (x^2 - 3)^{2/3}$ is increasing for which values of x?

(A) $-\sqrt{3} \le x \le \sqrt{3}$

(B) $x \le -\sqrt{3}$ or $x > \sqrt{3}$

(C) $-3 \le x \le 3$

(D) $-\sqrt{3} < x < 0$ or $x > \sqrt{3}$

(E) $f(x)$ is never increasing.

37. If $y = \tan(\arccos x)$, then $\dfrac{dy}{dx} =$

(A) $\dfrac{-1}{x^2\sqrt{1-x^2}}$

(B) $\dfrac{1}{x^2\sqrt{1-x^2}}$

(C) $\dfrac{-1}{x\sqrt{1-x^2}}$

(D) $\dfrac{1}{x\sqrt{1-x^2}}$

(E) $\dfrac{-1}{x^2\sqrt{x^2-1}}$

38. Which of the following statements is true?

(A) $\log_{\frac{1}{2}} 2 < \log_{\frac{1}{\sqrt{2}}} 2$

(B) $\log_3(2+4) = \log_3 2 + \log_3 3$

(C) $\log_{10} 2 > \log_{10} 4$

(D) $\log_{\frac{1}{5}}(5\sqrt{5}) = \dfrac{2}{3}$

(E) $\log_{\frac{1}{2}} 2 - \log_{\frac{1}{2}} 4 = \log_{\frac{1}{2}} 2$

39. Let $f(x) = (x^2 - 3)^2$. The local maximum of $f'(x)$ is

(A) 7.99

(B) 5.80

(C) 3.25

(D) −7.99

(E) 6.28

40. $f(x) = \dfrac{x^2}{e^x}$. $\displaystyle\int_0^5 \dfrac{x^2}{e^x}$ is

(A) 2.25

(D) 1.75

(B) 1.3

(E) 12.5

(C) 7.1

ADVANCED PLACEMENT
CALCULUS AB EXAM V

SECTION II

Time: 1 hour and 30 minutes
         6 problems

DIRECTIONS: Show all your work. Grading is based on the methods used to solve the problem as well as the accuracy of your final answers. Please make sure all procedures are clearly shown.

NOTES:
1. Unless otherwise specified, answers can be given in unsimplified form.

2. The domain of function f is assumed to be the set of all real numbers x for which $f(x)$ is a real number.

1.   Let f be the function given by $f(x) = \dfrac{x^2 - 4}{1 - x^2}$.

   (A) Find the domain of f.

   (B) Find the range of f.

   (C) Find the equations for each vertical and each horizontal asymptote.

   (D) Find the critical points.

2. At a child's party, there is a contest to see which child can run the fastest from a point 20 feet from a fence, to the fence, and then to a point 30 feet from the fence as in the figure shown. Find the point on the fence to where the children should run in order to minimize the distance by using the following:

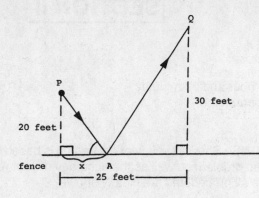

(A) Express $\overline{PA}$ and $\overline{QA}$ in terms of x.

(B) Find x.

3. The current I in a circuit is given by $I(t) = 20 \sin (311t) + 40 \cos (311t)$. Let $\cos \theta = \dfrac{1}{\sqrt{5}}$ and $\sin \theta = \dfrac{2}{\sqrt{5}}$.

(A) Verify, by use of the sum formula for the sine function, that $I(t) = 20\sqrt{5} \sin (311t + \theta)$.

(B) Determine the peak current (the maximum value of I) by the second derivative test.

4.   Use the information below to answer the following questions about $y = f(x)$:

$$\lim_{x \to 2^-} f(x) = -\infty, \quad \lim_{x \to 2^+} f(x) = \infty, \quad \lim_{x \to -1^-} f(x) = \infty$$

$$\lim_{x \to -1^+} f(x) = -\infty, \quad \lim_{x \to \infty} f(x) = 0, \quad \lim_{x \to -\infty} f(x) = 0$$

Interval	$(-\infty, -4)$	$(-4, 0)$	$(0, \infty)$
sign of $f'(x)$	$-$	$+$	$-$

$$f(-4) = -\frac{2}{3}, \quad f(0) = -6, \quad f(-2) = 0$$

$f(2)$ and $f(-1)$ are undefined.

(A)   Find the intervals where f is increasing and where f is decreasing.

(B)   Find the equation for each horizontal and each vertical asymptote.

(C)   Find the local maximum and minimum values of f.

(D)   Sketch f. Label asymptotes, local extrema, and intercepts.

5.	Let $f(x) = \frac{2}{3}x^{\frac{3}{2}}$ and suppose that the line $y = cx + d$ is tangent to $f(x)$ at x_0.

(A)	If $x_0 = 2$, find c and d.

(B)	If $c = 1$, find x_0 and d.

(C)	Find $\int_0^3 \frac{2}{3}x^{\frac{3}{2}}$.

6.	A projectile is fired directly upward from the ground with an initial velocity of 112 feet/second, and its distance above the ground after t seconds is $s(t) = 112t - 16t^2$ feet.

(A)	What are the velocity and acceleration of the projectile at $t = 3$ seconds?

(B)	At what time does the projectile reach its maximum height?

(C)	What is the velocity at the moment of impact?

(D)	The length of the path of the projectile is approximately 392 feet. If a bug traverses the arc of the curve
$$y = \frac{\sqrt{(x^2 - 2)^3}}{3}$$
from $x = 1$ foot to $x = 10$ feet, which distance is longer, the path of the projectile or the path traversed by the bug?

ADVANCED PLACEMENT
CALCULUS AB
EXAM V

ANSWER KEY

Section I

1.	B		21.	A
2.	C		22.	D
3.	A		23.	C
4.	E		24.	A
5.	D		25.	D
6.	D		26.	E
7.	B		27.	B
8.	A		28.	A
9.	D		29.	B
10.	D		30.	D
11.	A		31.	C
12.	E		32.	D
13.	A		33.	A
14.	C		34.	D
15.	C		35.	E
16.	B		36.	D
17.	A		37.	A
18.	C		38.	B
19.	C		39.	A
20.	B		40.	D

Section II

See Detailed Explanations of Answers.

ADVANCED PLACEMENT
CALCULUS AB EXAM V

SECTION I

DETAILED EXPLANATIONS
OF ANSWERS

1.   **(B)**

$$\frac{d}{dx}\left(3x^2 - x^2y^3 + 4y\right) = \frac{d}{dx}(12)$$

$$6x - 2xy^3 - x^2 3y^2 y' + 4y' = 0$$

$$6x - 2xy^3 + y'\left(-3x^2y^2 + 4\right) = 0$$

$$y'\left(-3x^2y^2 + 4\right) = -6x + 2xy^3$$

$$y' = \frac{-6x + 2xy^3}{-3x^2y^2 + 4}$$

$$\frac{dy}{dx} = \frac{-6x + 2xy^3}{-3x^2y^2 + 4}$$

2.   (C)

$$\int_1^4 \frac{5x^2 - x}{2\sqrt{x}}\, dx$$

$$= \int_1^4 \frac{5x^2 - x}{2x^{1/2}}\, dx$$

$$= \int_1^4 \left(\frac{5x^{3/2}}{2} - \frac{x^{1/2}}{2} \right) dx$$

$$= \left(x^{5/2} - \frac{1}{3}\, x^{3/2} \right) \Big|_1^4$$

$$= \left(2^5 - 1^5 \right) - \frac{1}{3} \left(2^3 - 1^3 \right)$$

$$= (31) - \frac{7}{3} = \frac{86}{3}$$

3.   (A)   (i) $A = \displaystyle\int_0^2 \left(4y - y^3 \right) dy = \left(2y^2 - \frac{y^4}{4} \right) \Big|_0^2$

$$= 2(2)^2 - \frac{1}{4}(2)^4$$

$$= 8 - 4$$

$$= 4$$

or (ii) $A = \displaystyle\int_0^8 \left(\sqrt[3]{x} - \frac{x}{4} \right) dx = \int_0^8 \left(x^{1/3} - \frac{1}{4}x \right) dx$

$$= \left(\frac{3}{4}\, x^{4/3} - \frac{1}{8}\, x^2 \right) \Big|_0^8$$

$$= \frac{3}{4}(8)^{4/3} - \frac{1}{8}(8)^2$$

$$= \frac{3}{4}(16) - 8$$

$$= 4$$

Note: Method (i) is simpler since it does not require solving for y or integration involving fractional exponents.

4.   (E)

$f(x)$ is continuous for $x > -1$ and $x < -1$ since polynomials are continuous for all reals. We must determine if $f(x)$ is continuous at $x = -1$

(i) $f(-1) = 3(-1)^2 + 2 = 3 + 5$, thus $f(-1)$ is defined.

(ii) $\lim\limits_{x \to -1} f(x)$ exists if $\lim\limits_{x \to -1^+} f(x) = \lim\limits_{x \to -1^-} f(x)$.

$$\lim\limits_{x \to -1^+} f(x) = \lim\limits_{x \to -1^+} (3x^2 + 2)$$

$$= 3(-1)^2 + 2$$

$$= 5$$

Therefore $\lim\limits_{x \to -1^-} f(x)$ must also equal 5.

$$\lim\limits_{x \to -1^-} (-cx + 5) = 5$$

$$\Rightarrow -c(-1) + 5 = 5 \Rightarrow c = 0$$

Thus, for $c = 0$

$\lim\limits_{x \to -1^+} f(x) = \lim\limits_{x \to -1^-} f(x)$, therefore $\lim\limits_{x \to -1} f(x)$ exists .

Since $\lim\limits_{x \to -1} f(x) = f(-1) = 5$, $f(x)$ is continuous.

The graph of $f(x)$ with $c = 0$ is sketched below:

$$f(x) = \begin{cases} 3x^2 + 2 & x \geq -1 \\ 5 & x < -1 \end{cases}$$

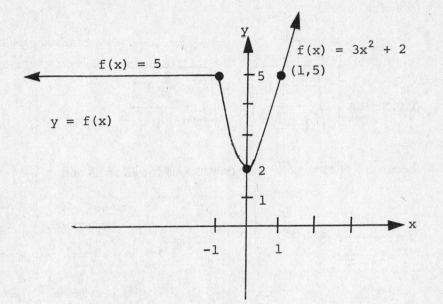

5. (D)

$$\lim_{x \to \infty} \sqrt[3]{\frac{8 + x^2}{x(x+1)}} = \lim_{x \to \infty} \sqrt[3]{\frac{8 + x^2}{x^2 + x}}$$

$$= \lim_{x \to \infty} \sqrt[3]{\frac{(8 + x^2) \dfrac{1}{x^2}}{(x^2 + x) \dfrac{1}{x^2}}}$$

$$= \lim_{x \to \infty} \sqrt[3]{\frac{\left(\dfrac{8}{x^2}\right) + 1}{1 + \dfrac{1}{x}}}$$

$$= \sqrt[3]{\lim_{x \to \infty} \frac{\left(\dfrac{8}{x^2}\right) + 1}{1 + \dfrac{1}{x}}}$$

237

$$= \sqrt[3]{\frac{0+1}{1+0}}$$

$$= \sqrt[3]{1}$$

$$= 1$$

Note: $\lim\limits_{x\to\infty} \left[\sqrt[3]{\dfrac{\dfrac{8}{x^2}+1}{1+\dfrac{1}{x}}} \right] = \sqrt[3]{\lim\limits_{x\to\infty} \dfrac{\dfrac{8}{x^2}+1}{1+\dfrac{1}{x}}}$

because $f(x) = \sqrt[3]{x}$ is continuous for all reals and

$\lim\limits_{x\to\infty} \dfrac{\dfrac{8}{x^2}+1}{1+\dfrac{1}{x}}$ exists.

6. (D)

Let $\theta = \text{arcsec} x$.

By the identity $\tan^2\theta + 1 = \sec^2\theta$, we have

$\tan^2(\text{arcsec} x) + 1 = \sec^2(\text{arcsec} x)$

$\Rightarrow y^2 + 1 = x^2$

$\Rightarrow y = \sqrt{x^2 - 1}$

Hence $y = (x^2 - 1)^{1/2}$ and

$$\frac{dy}{dx} = \frac{2x}{2} (x^2 - 1)^{-1/2}$$

$$= \frac{x}{\sqrt{x^2 - 1}} .$$

7.   (B)

Since $g(x) = f^{-1}(x)$ then $f(g(x)) = x$

Differentiating both sides with respect to x gives:

$$\frac{d}{dx} f(g(x)) = \frac{d}{dx}(x)$$

Using the chain rule, we obtain $f'(g(x)) \cdot g'(x) = 1$

Hence, $g'(x) = \dfrac{1}{f'(g(x))}$

$$g'(4) = \frac{1}{f'(g(4))} \qquad \text{since } f(1) = 4 \text{ then } g(4) = 1$$

$$= \frac{1}{f'(1)} \qquad\qquad \text{since } f(x) = 10x^4 - 3x^2 + 2x$$

$$\qquad\qquad\qquad\qquad f'(1) = 10 - 3 + 2$$

$$= \frac{1}{9} \qquad\qquad\qquad\qquad = 9$$

8.   (A)

$$\lim_{x \to 0^-} (1 - x)^{2/x}$$

Note: x approaches zero from the left since $(1 - x)^{2/x}$ is not defined for every value of $x > 1$.

Let $y = (1 - x)^{2/x}$

$$\ln y = \ln(1 - x)^{2/x}$$

$$= \frac{2}{x} \ln(1 - x)$$

$$= \frac{2 \ln(1 - x)}{x}$$

$$\lim_{x\to 0^-}\ \ln y = \lim_{x\to 0^-}\ \frac{2\ln(1-x)}{x} \quad \text{by L'Hopital's Rule,}$$

$$= \lim_{x\to 0^-}\ \frac{2\left(\dfrac{1}{1-x}\right)(-1)}{1}$$

$$= \lim_{x\to 0^-}\ -\frac{2}{1-x}$$

$$= -2$$

Therefore, $\displaystyle\lim_{x\to 0^-}\ \ln y = -2$

$$\lim_{x\to 0^-}\ (1-x)^{2/x} = \lim_{x\to 0^-}\ y$$

$$= \lim_{x\to 0^-}\ e^{\ln y}$$

$$= e^{\lim_{x\to 0^-}\ \ln y}$$

$$= e^{-2}$$

9. (D)

We know that $f(x) = |x| = \begin{cases} x & \text{if } x \geq 0 \\ -x & \text{if } x < 0 \end{cases}$

Hence, $f'(x) = \begin{cases} 1 & \text{if } x \geq 0 \\ -1 & \text{if } x < 0 \end{cases} = \dfrac{x}{|x|}$, for $x \neq 0$

Therefore, $f'(x)$ is a real number for all $x \neq 0$.

10. (D)

We note $1° = \dfrac{\pi}{180}$ radians, hence $x° = x \cdot \dfrac{\pi}{180}$ radians.

$$\frac{d}{dx}\,(\sin x°) = \frac{d}{dx}\sin\left(x \cdot \frac{\pi}{180}\right)$$

$$= \frac{\pi}{180} \cos \left(x \cdot \frac{\pi}{180} \right)$$

$$= \frac{\pi}{180} \cos (x°)$$

11. (A)

$$\int \frac{\log (x^3 \cdot 10^x)}{x} \, dx = \int \frac{\log x^3 + \log 10^x}{x} \, dx$$

$$= \int \frac{3 \log x + x}{x} \, dx$$

$$= \int \left(\frac{3 \log x}{x} + 1 \right) dx$$

Let $u = \log x = \frac{\ln x}{\ln 10}$, $du = \frac{dx}{x \ln 10}$

$$3 \ln 10 \int \frac{\log x}{x \ln 10} \, dx + \int 1 dx = 3 \ln 10 \int u \, du + x + C$$

$$= 3 \ln 10 \frac{u^2}{2} + x + C'$$

$$= \frac{3 \ln 10 \, (\log x)^2}{2} + x + C'$$

12. (E)

We see that $f'(x) = 6x^2 - 18x + 12 = 0$

$$\Rightarrow x^2 - 3x + 2 = 0$$

$$\Rightarrow (x - 1)(x - 2) = 0$$

$$\Rightarrow x = 1, 2 .$$

So the invervals to consider are $x < 1$, $1 < x < 2$ and $x > 2$.

We try values of x in each interval to see whether $f(x)$ is increasing $(f'(x) > 0)$ or decreasing $(f'(x) < 0)$ there.

241

For $x < 1$, we can try $x = 0$.

Then $f'(x) = f'(0) = 12 > 0$, so $f(x)$ is increasing when $x < 1$.

We now try $x = 1.5$ for the second interval.

$f'(1.5) = 6(1.5)^2 - 18(1.5) + 12 = 6(2.25) - 27 + 12 = -1.5 < 0$

so $f(x)$ is decreasing on $1 < x < 2$.

Now we try $x = 3$ for the inverval $x > 2$. Then

$f'(3) = 54 - 54 + 12 = 12 > 0$, so $f(x)$ is increasing for $x > 2$.

The interval on which $f(x)$ is decreasing is therefore $1 < x < 2$.

13. (A)

$$\lim_{x \to +\infty} \frac{|x|}{|x+1|+x} = \lim_{x \to +\infty} \frac{x}{x+1+x}$$

$$= \lim_{x \to +\infty} \left(\frac{x}{2x+1}\right)\left(\frac{\frac{1}{x}}{\frac{1}{x}}\right)$$

$$= \lim_{x \to +\infty} \frac{1}{2 + \frac{1}{x}}$$

$$= \frac{1}{2}$$

$$\lim_{x \to -\infty} \frac{|x|}{|x+1|+x} = \lim_{x \to -\infty} \frac{-x}{-(x+1)+x}]$$

$$= \lim_{x \to -\infty} \frac{-x}{-1}$$

$$= \lim_{x \to -\infty} x$$

$$= -\infty$$

Thus $y = \frac{1}{2}$ is a horizontal asymptote.

14. (C)

We want to minimize the distance $D = \sqrt{(x-6)^2 + (y-3)^2}$ from a point (x, y) to the point $(6, 3)$.

We can do this more easily if we square both sides to remove the square root sign.

We have $D^2 = (x-6)^2 + (y-3)^2$.

We can now replace y with x^2 since the point is on the parabola $y = x^2$.

We expand to get:
$$D^2 = (x-6)^2 + (x^2-3)^2$$
$$= x^2 - 12x + 36 + x^4 - 6x^2 + 9$$
$$= x^4 - 5x^2 - 12x + 45.$$

We now take the derivative and set it equal to zero:
$$4x^3 - 10x - 12 = 0.$$

Dividing by 2 gives $0 = 2x^3 - 5x - 6$

$$0 = (x-2)(2x^2 + 4x + 3)$$

$$\Rightarrow x = 2 \text{ or } 2x^2 + 4x + 3 = 0$$

$$\Rightarrow x = \frac{-4 \pm \sqrt{16 - 24}}{4},$$

which is not a real number.

So $x = 2$ and $y = x^2 = 4$. The point on the parabola is then $(2, 4)$.

15. (C)

$$\lim_{x \to -1} \frac{\left(\sqrt{x^2+3} - 2\right)\left(\sqrt{x^2+3} + 2\right)}{(x+1)\left(\sqrt{x^2+3} + 2\right)}$$

$$= \lim_{x \to -1} \frac{x^2 + 3 - 4}{(x+1)\left(\sqrt{x^2+3} + 2\right)}$$

$$= \lim_{x \to -1} \frac{x^2 - 1}{(x+1)\left(\sqrt{x^2+3}+2\right)}$$

$$= \lim_{x \to -1} \frac{(x-1)(x+1)}{(x+1)\left(\sqrt{x^2+3}+2\right)}$$

$$= \lim_{x \to -1} \frac{x-1}{\sqrt{x^2+3}+2}$$

$$= \left(-\frac{2}{2+2}\right)$$

$$= -\frac{1}{2}$$

Alternately, apply L'Hôpital's rule.

$$\lim_{x \to -1} \frac{\sqrt{x^2+3}-2}{x+1}$$

$$= \lim_{x \to -1} \frac{\frac{1}{2}(x^2+3)^{-1/2}(2x)}{1}$$

$$= \lim_{x \to -1} \frac{x}{\sqrt{x^2+3}} = -\frac{1}{\sqrt{4}}$$

$$= -\frac{1}{2}$$

16. (B)

$$y^2 = 2x - 2 \qquad y = x - 5$$

To find the points of intersection, replace y with $x - 5$ in

$$y^2 = 2x - 2$$

$$(x-5)^2 = 2x - 2$$

$$x^2 - 10x + 25 = 2x - 2$$

$$x^2 - 12x + 27 = 0$$

$$(x-9)(x-3) = 0$$

$$x = 9 \Rightarrow y = 9 - 5 = 4, \ (x,y) = (9,4)$$

$$x = 3 \Rightarrow y = 3 - 5 = -2, \ (x,y) = (3,-2)$$

The graph shows curves $x = \dfrac{y^2 + 2}{2}$ and $x = y + 5$ with points $(9, 4)$, $(3, -2)$.

$$A = \int_{-2}^{4} \left[(y + 5) - \left(\frac{y^2 + 2}{2} \right) \right] dy$$

$$= \frac{1}{2} \int_{-2}^{4} (2y + 10 - y^2 - 2) \; dy$$

$$= \frac{1}{2} \int_{-2}^{4} (-y^2 + 2y + 8) \; dy$$

$$= \frac{1}{2} \left[-\frac{y^3}{3} + y^2 + 8y \right] \Big|_{-2}^{4}$$

$$= \frac{1}{2} \left[-\frac{1}{3} (64 + 8) + (16 - 4) + 8 (4 + 2) \right]$$

$$= \frac{1}{2} \left[-\frac{72}{3} + 12 + 48 \right]$$

$$= \frac{1}{2} [36]$$

$$= 18$$

17.   (A)

$$y = t^2 e^{3t^2 + \sqrt{t}} = t^2 (e^{3t^2 + t^{1/2}})$$

$$y' = 2t \, (e^{3t^2 + t^{1/2}}) + (e^{3t^2 + t^{1/2}})(6t + \frac{1}{2} t^{-1/2}) \cdot t^2$$

$$y' = 2(1)(e^{3+1}) + e^{3+1}(6 + \frac{1}{2} (1)) \cdot 1$$

$$= 2e^4 + 6.5e^4$$

$$= 8.5e^4 \approx 464.084$$

18.   (C)

$$\int \frac{e^{2x}}{e^x - 3} \, dx \qquad\qquad \text{Let } u = e^x - 3 \Rightarrow u + 3 = e^x$$
$$\qquad\qquad\qquad\qquad du = e^x \, dx$$

$$\int \frac{e^x(e^x dx)}{e^x - 3} = \int \frac{(u + 3)}{u} \, du$$

$$= \int \left(1 + \frac{3}{u}\right) du$$

$$= u + 3 \ln |u| + C$$

$$= (e^x - 3) + 3 \ln|e^x - 3| + C$$

19.   (C)

$$xy - x = 2y - 5$$

$$xy - 2y = x - 5$$

$$y(x - 2) = x - 5$$

$$y = \frac{x - 5}{x - 2}.$$

y is a function of x with domain $\{x \mid x \neq 2\}$.

To determine the range we find the inverse function:

$$x = \frac{y - 5}{y - 2}$$

$$xy - 2x = y - 5$$

$$xy - y = 2x - 5$$

$$y(x - 1) = 2x - 5$$

$$y = \frac{2x - 5}{x - 1}$$

$$f^{-1}(x) = \frac{2x - 5}{x - 1}$$

The domain of $f^{-1}(x)$ is $\{x \mid x \neq 1\}$. Thus the range of $f(x) = \frac{x - 5}{x - 2}$ is $\{x \mid x \neq 1\}$.

Hence only I and II are true.

20.   (B)

$h(x) = \sqrt{x^2 - 4}$, $x \leq -2$. Let $y = \sqrt{x^2 - 4}$.

To find the inverse, let $x = \sqrt{y^2 - 4}$ (interchange x and y)

$$\Rightarrow x^2 = y^2 - 4$$

$$\Rightarrow x^2 + 4 = y^2$$

$$\pm \sqrt{x^2 + 4} = y.$$

Since $x \leq -2$ and we interchanged x and y, then $y \leq -2$. It follows that $y = -\sqrt{x^2 + 4}$ or $h^{-1}(x) = -\sqrt{x^2 + 4}$.

21.   (A)

$$y = 1 + 2^{x+3}$$

$$y' = 2^{x+3} \ln 2$$

Since $y' > 0$, y is always increasing. Graphs (A), (B) and (D) are increasing. The ordered pair $(-3, 2)$ is a point on the graph, and (A) is the only one of these graphs which includes $(-3, 2)$.

22.   (D)

$$\lim_{x \to 0} \frac{\sin 2x}{x \cos x} = \lim_{x \to 0} \frac{2 \sin x \cos x}{x \cos x}$$

$$= 2 \lim_{x \to 0} \frac{\sin x}{x} . \quad \text{Apply L'Hôpital's rule:}$$

$$= 2 \lim_{x \to 0} \frac{\cos x}{1}$$

$$= 2 \, (1)$$

$$= 2$$

Note: $\lim_{x \to 0} \dfrac{\sin x}{x} = 1$ is also a well-known theorem.

23.   (C)

We write $xy = 4$ as $y = \dfrac{4}{x}$. We want to find the area between $y = \dfrac{4}{x}$ and $y = 0$ (the x-axis), with $x = e$ and $x = 2e$ as the limits of integration.

$$\text{Hence} \quad A = \int_e^{2e} \left(\frac{4}{x} - 0 \right) dx = 4 \int_e^{2e} \frac{dx}{x}$$

$$= 4 \ln x \, \Big|_e^{2e}$$

$$= 4 \, (\ln 2e - \ln e).$$

But $\ln 2e = \ln 2 + \ln e$, so $4(\ln 2e - \ln e)$

$$= 4(\ln 2 + \ln e - \ln e)$$

$$= 4\ln 2.$$

$$= \approx 2.773$$

24.   (A)

$$\frac{\cos 2\beta - \sin 2\beta}{\sin \beta \cos \beta} = \frac{(\cos^2 \beta - \sin^2 \beta) - (2\sin \beta \cos \beta)}{\sin \beta \cos \beta}$$

$$= \frac{\cos^2 \beta}{\sin \beta \cos \beta} - \frac{\sin^2 \beta}{\sin \beta \cos \beta} - \frac{2\sin \beta \cos \beta}{\sin \beta \cos \beta}$$

$$= \frac{\cos \beta}{\sin \beta} - \frac{\sin \beta}{\cos \beta} - 2$$

$$= \cot \beta - \tan \beta - 2$$

25.   (D)

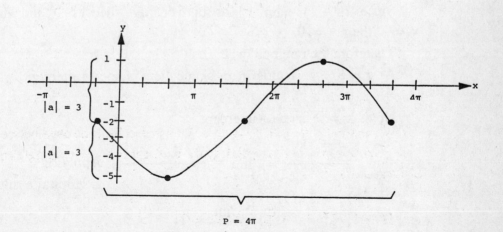

$$P = 4\pi$$

$$P = \frac{11\pi}{3} - \left(-\frac{\pi}{3}\right)$$

$$= \frac{12\pi}{3}$$

$$= 4\pi$$

$$= \frac{2\pi}{b}$$

$$b = \frac{2\pi}{4\pi} = \frac{1}{2}$$

$|a| = 3$ $\left(-\frac{\pi}{3}, -2\right)$ is the initial point on the graph.

Since the graph opens downward, $a = -3$.

$$y + 2 = -3\sin\frac{1}{2}\left(x + \frac{\pi}{3}\right)$$

$$\therefore f(x) = -3\sin\frac{1}{2}\left(x + \frac{\pi}{3}\right) - 2$$

26.   (E)

$$y = \frac{2(x-1)^2}{x^2}$$

Since $(x-1)^2$ and x^2 are both positive, then $y > 0$, and when $x = 1$ then $y = 0$.

Therefore the range is $y \geq 0$.

At $x = 0$, y is undefined, hence there is no y–intercept.

Note: $x = 1$ is an x–intercept.

$$\lim_{x \to \pm\infty} \frac{2(x-1)^2}{x^2} = \lim_{x \to \pm\infty} \frac{2 \cdot 2(x-1)}{2x}, \quad \begin{array}{l}\text{Applying} \\ \text{L'Hôpital's rule,}\end{array}$$

$$= 2 \lim_{x \to \pm\infty} \frac{1}{1}$$

$$= 2$$

Therefore $y = 2$ is a horizontal asymptote. Only I and III are true.

27. (B)

$g(x + 3) = x^2 + 2$.

Since $x^2 + 2$ does not involve the expression $(x + 3)$, we first subtract 3 from x. If $g(x) = x - 3$ then $g(x + 3) = x + 3 - 3 = x$ which is not $x^2 + 2$. But if we square $(x - 3)$ and add 2 we have

$$g(x) = (x - 3)^2 + 2 \quad \text{then}$$

$$g(x + 3) = (x + 3 - 3)^2 + 2$$

$$= x^2 + 2.$$

Therefore $g(x) = (x - 3)^2 + 2$.

28. (A)

$$y = \frac{x^2}{\sqrt[3]{3x^2 + 1}} = \frac{x^2}{(3x^2 + 1)^{1/3}}$$

$$y' = \frac{2x(3x^2 + 1)^{1/3} - \frac{1}{3}(3x^2 + 1)^{-2/3}(6x) \cdot x^2}{(3x^2 + 1)^{2/3}}$$

$$= \frac{2x(3x^2 + 1)^{1/3} - (3x^2 + 1)^{-2/3} \, 2x^3}{(3x^2 + 1)^{2/3}}$$

$$y'(1) = \frac{2(3 + 1)^{1/3} - (3 + 1)^{-2/3}(2)}{(3 + 1)^{2/3}}$$

$$= \frac{2(4)^{1/3} - 2(4)^{-2/3}}{(4)^{2/3}} \cdot \frac{4^{2/3}}{4^{2/3}}$$

$$= \frac{2(4) - 2}{(4)^{4/3}}$$

$$= \frac{6}{4(4)^{1/3}}$$

$$= \frac{3}{2\sqrt[3]{4}} \cdot \frac{\sqrt[3]{2}}{\sqrt[3]{2}}$$

$$= \frac{3\sqrt[3]{2}}{2\sqrt[3]{8}}$$

$$= \frac{3\sqrt[3]{2}}{4}$$

Therefore the slope of the tangent line at $x = 1$ is $\dfrac{3\sqrt[3]{2}}{4}$.

Calculator: $2\,^3\sqrt{x} \times .75 = \approx 0.945$

29.　(B)

$x^4 + xy + y^4 = 1$.

Differentiate both sides with respect to t.

$$\frac{d}{dt}\left(x^4 + xy + y^4\right) = \frac{d}{dt}(1)$$

$$4x^3\frac{dx}{dt} + \frac{dx}{dt} \cdot y + \frac{dy}{dt} \cdot x + 4y^3\frac{dy}{dt} = 0$$

$$\frac{dy}{dt}\left(x + 4y^3\right) = -\left(4x^3 + y\right)\frac{dx}{dt}$$

$$\frac{dy}{dt} = -\left(\frac{4x^3 + y}{x + 4y^3}\right)\frac{dx}{dt}.$$

30.　(D)

$$g(u) = \sqrt{u^3 + 2}, \quad g'(u) = \frac{1}{2}\left(u^3 + 2\right)^{-1/2} \cdot 3u^2$$

$$\frac{d}{dx}\left(g(f(x))\right) = g'(f(x)) \cdot f'(x)$$

At $x = 1$

$$= g'(f(1)) \cdot f'(1)$$

$$= g'(2) \cdot (-5)$$

$$= \frac{1}{2}\left(2^3 + 2\right)^{-1/2} \cdot 3(2^2) \cdot (-5)$$

$$= \frac{1}{2\sqrt{10}} \cdot 12 \cdot (-5)$$

252

$$= -\frac{30}{\sqrt{10}} \frac{\sqrt{10}}{\sqrt{10}}$$

$$= -\frac{30\sqrt{10}}{10}$$

$$= -3\sqrt{10} \approx -9.487$$

Calculator: $+/- 3 \times 10 \sqrt{x} = \approx -9.487$

31.   (C)
$$y = \frac{1}{\sin(t + \sqrt{t})} = \left(\sin\left(t + t^{1/2}\right)\right)^{-1}$$

$$y' = -1 \left(\sin\left(t + t^{1/2}\right)\right)^{-2} \cdot \cos(t + t^{1/2}) \cdot \left(1 + \frac{1}{2} t^{-1/2}\right)$$

$$y'(1) = -1(\sin 2)^{-2} \cdot \cos 2 \cdot \left(1 + \frac{1}{2}\right)$$

$$= -\frac{3}{2} \frac{\cos 2}{\sin^2 2} \approx -1230.802$$

Calculator: $2 \cos \div (2 \sin x^2) = x +/- 1.5 = \approx -1230.802$

32.   (D)
Let $u = \arctan x$, $dv = dx$. Then $du = \dfrac{1}{1 + x^2} dx$ and $v = x$, so integration by parts gives:

$$\int \arctan x \, dx = x \arctan x - \int \frac{x}{1 + x^2} \, dx.$$

Now let $z = 1 + x^2$, $dz = 2x \, dx$.

Then $\displaystyle\int \frac{x}{1 + x^2} \, dx = \frac{1}{2} \int \frac{dz}{z}$

$$= \frac{1}{2} \ln z + C$$

$$= \frac{1}{2} \ln(1 + x^2) + C.$$

So $\displaystyle\int \arctan x \, dx = x \arctan x - \frac{1}{2} \ln(1 + x^2) + C.$

33. (A)

$$y = -\ln\left|\frac{1 + \sqrt{1 - x^2}}{x}\right| = \ln\left|\frac{x}{1 + \sqrt{1 - x^2}}\right|$$

$$= \ln|x| - \ln|1 + \sqrt{1 - x^2}|$$

$$\frac{dy}{dx} = \frac{1}{x} - \frac{1}{1 + \sqrt{1 - x^2}} \cdot \frac{1}{2}(1 - x^2)^{-1/2}(-2x)$$

$$= \frac{1}{x} + \frac{1}{(1 + \sqrt{1 - x^2})} \frac{x}{(\sqrt{1 - x^2})}$$

$$= \frac{(1 + \sqrt{1 - x^2})\sqrt{1 - x^2} + x^2}{x(1 + \sqrt{1 - x^2})(\sqrt{1 - x^2})}$$

$$= \frac{\sqrt{1 - x^2} + (1 - x^2) + x^2}{x(1 + \sqrt{1 - x^2})(\sqrt{1 - x^2})}$$

$$= \frac{1 + \sqrt{1 - x^2}}{x(1 + \sqrt{1 - x^2})(\sqrt{1 - x^2})}$$

$$= \frac{1}{x\sqrt{1 - x^2}}$$

34. (D)

$$F = \frac{6000\,k}{k\sin\theta + \cos\theta} = 6000\,k(k\sin\theta + \cos\theta)^{-1}$$

$$\frac{dF}{d\theta} = -6000\,k(k\sin\theta + \cos\theta)^{-2}(k\cos\theta - \sin\theta)$$

$$= -\frac{6000\,k(k\cos\theta - \sin\theta)}{(k\sin\theta + \cos\theta)^2}$$

$$\frac{dF}{d\theta} = 0 \implies k\cos\theta - \sin\theta = 0 \implies k = \tan\theta$$

Hence $\theta = \arctan k$.

254

35.   (E)

This problem can be solved directly by using your calculator. For example,

$$fnInt \left(\frac{1}{\sin (x + \sqrt{x})}, x, 0, 1 \right)$$

gives 1.65 .

36.   (D)

$$f(x) = (x^2 - 3)^{\frac{2}{3}}$$

$$f'(x) = \frac{2}{3} (x^2 - 3)^{-\frac{1}{3}} \cdot 2x$$

$$= \frac{4x}{3} (x^2 - 3)^{-\frac{1}{3}}$$

$f'(x) > 0$ for i) $x > 0$ and $x^2 - 3 > 0$ or

ii) $x < 0$ and $x^2 - 3 < 0$

i) $x > 0$ and $x^2 - 3 > 0$

$\Rightarrow$ $x > 0$ and $x < -\sqrt{3}$ or $x > \sqrt{3}$

so $x > \sqrt{3}$

ii) $x < 0$ and $x^2 - 3 < 0 \Rightarrow x < 0$ and $x^2 < 3$

$x < 0$ and $-\sqrt{3} < x < \sqrt{3}$

$-\sqrt{3} < x < 0$

Thus $f(x) = (x^2 - 3)^{\frac{2}{3}}$ is increasing for x such that

$$-\sqrt{3} < x < 0 \quad \text{or} \quad x > \sqrt{3}.$$

37.   (A)

Let $\theta = \arccos x$, $x = \cos\theta$, $\dfrac{1}{x} = \sec\theta$

$$y = \tan(\arccos x) = \tan\theta.$$

By the identity $\tan^2\theta + 1 = \sec^2\theta$, we have

$$y^2 + 1 = \frac{1}{x^2} \Rightarrow y^2 = \frac{1}{x^2} - 1$$

$$\Rightarrow y^2 = \frac{1 - x^2}{x^2}$$

$$\Rightarrow y = \frac{\sqrt{1 - x^2}}{x}$$

$$\frac{dy}{dx} = \frac{d}{dx}\left(\frac{\sqrt{1 - x^2}}{x}\right)$$

$$= \frac{x\left[\frac{1}{2}(1 - x^2)^{-\frac{1}{2}}(-2x)\right] - \sqrt{1 - x^2}\,(1)}{x^2}$$

$$= \frac{-x^2(1 - x^2)^{-\frac{1}{2}} - (1 - x^2)^{-\frac{1}{2}}}{x^2} \cdot \left[\frac{(1 - x^2)^{\frac{1}{2}}}{(1 - x^2)^{\frac{1}{2}}}\right]$$

$$= \frac{-x^2 - (1 - x^2)}{x^2(1 - x^2)^{\frac{1}{2}}}$$

$$= \frac{-1}{x^2\sqrt{1 - x^2}}$$

38.   (B)

(A) $\log_{\frac{1}{2}} 2 < \log_{\frac{1}{\sqrt{2}}} 2$

$-1 < -2$ False

(C) $\log_{10} 2 > \log_{10} 4$

$\log_{10} 2 > \log_{10} 2^2$

$\log_{10} 2 > 2\log_{10} 2$

$1 > 2$ False

(D) $\log_{\frac{1}{5}} 5\sqrt{5} = \frac{2}{3}$

$\log_{\frac{1}{5}} 5 \cdot 5^{\frac{1}{2}} = \frac{2}{3}$

$\log_{\frac{1}{5}} 5^{\frac{3}{2}} = \frac{2}{3}$

$\log_{\frac{1}{5}} \left(\frac{1}{5}\right)^{-\frac{3}{2}} = \frac{2}{3}$

$-\frac{3}{2} = \frac{2}{3}$ False

(E) $\log_{\frac{1}{2}} 2 - \log_{\frac{1}{2}} 4 = \log_{\frac{1}{2}} 2$

$\log_{\frac{1}{2}} \frac{2}{4} = -1$

$\log_{\frac{1}{2}} \frac{1}{2} = -1$

$1 = -1$ False

(B) $\log_3 (2 + 4) = \log_3 2 + \log_3 3$

$\log_3 6 = \log_3 (2 \cdot 3)$

$\log_3 6 = \log_3 6$ True

257

39.　(A)
　　Draw the graphs of both $f(x)$ and $f'(x)$.

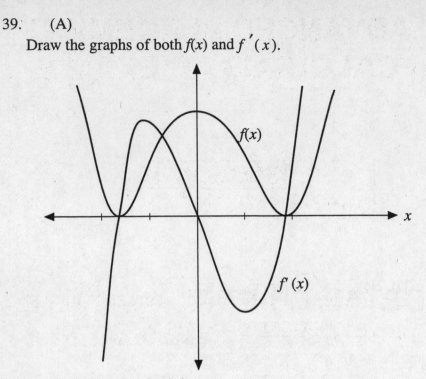

There is only one local minimum.
　　By tracing the graph of $f(x)$ to the minimum point, you can find
$f'_{max}(x) = 7.99$.

40.　(D)
　　You can solve this problem by directly using your calculator. For
example,

$$fnInt\left(\frac{x\,^{\wedge}2}{e\,^{\wedge}x}, x\,, 0, 5\right)$$

which would give 1.75.

ADVANCED PLACEMENT CALCULUS AB EXAM V

SECTION II

DETAILED EXPLANATIONS OF ANSWERS

1. (A)
    Domain $(f) = \{ x \mid x \neq \pm 1 \}$ because $f(x)$ is undefined when the denominator $1 - x^2 = 0$

$$\Rightarrow x^2 = 1$$
$$\Rightarrow x = \pm 1.$$

   (B)
    Range $(f) =$ Domain (f^{-1}). To find $f^{-1}(x)$, interchange x and y, then solve for y. We have:
$$x = \frac{y^2 - 4}{1 - y^2}$$

$$x(1 - y^2) = y^2 - 4$$

$$x - xy^2 = y^2 - 4$$

$$x + 4 = y^2 + xy^2 = y^2 (1 + x)$$

$$\frac{x+4}{1+x} = y^2$$

$$\pm \sqrt{\frac{x+4}{x+1}} = y = f^{-1}(x) .$$

Domain $(f^{-1}) = \{ x \mid x \neq -1 \}$, so Range $(f) = \{ y \mid y \neq -1 \}$.

(C)

We have vertical asymptotes where $f(x)$ is undefined, at $x = 1$ and $x = -1$.

We have horizontal asymptotes at $y = \lim\limits_{x \to \pm\infty} f(x)$.

We see $\lim\limits_{x \to \pm\infty} f(x) = \lim\limits_{x \to \pm\infty} \dfrac{x^2 - 4}{1 - x^2}$

$$= \lim_{x \to \pm\infty} \frac{1 - \dfrac{4}{x^2}}{\left(\dfrac{1}{x^2}\right) - 1}$$

$$= -\frac{1}{1}$$

$$= -1 ,$$

so $y = -1$ is the only horizontal asymptote.

(D)

The critical points occur where $f'(x) = 0$.

$$f'(x) = \frac{(1 - x^2)(2x) - (x^2 - 4)(-2x)}{(1 - x^2)^2} = 0$$

The numerator must be zero, so we have

$$2x - 2x^3 + 2x^3 - 8x = 0$$

$$-6x = 0$$

$$x = 0 \qquad \text{is the only critical point.}$$

Note: $f'(x)$ is undefined at $x = \pm 1$, but $x = \pm 1$ are not in the domain of f.

260

2.   (A)

$\overline{PA}^2 = 20^2 + x^2$ by the Pythagorean theorem.

$\overline{PA} = \sqrt{400 + x^2}$

Similarly, $\overline{QA} = \sqrt{30^2 + (25 - x)^2}$

$\qquad = \sqrt{900 + 625 - 50x + x^2}$

$\qquad = \sqrt{1525 - 50x + x^2}$

   (B)

We want to find x which minimizes $\overline{PA} + \overline{QA}$.

Let $f(x) = \overline{PA} + \overline{QA}$

$\qquad = \sqrt{400 + x^2} + \sqrt{1525 - 50x + x^2}$

Then $f'(x)$

$= \dfrac{1}{2}(400 + x^2)^{-1/2}2x + \dfrac{1}{2}(1525 - 50x + x^2)^{-1/2}(-50 + 2x)$

$= \dfrac{x}{\sqrt{400 + x^2}} + \dfrac{x - 25}{\sqrt{1525 - 50x + x^2}}$

$= \dfrac{x}{\sqrt{400 + x^2}} + \dfrac{x - 25}{\sqrt{900 + (25 - x)^2}}$

To minimize, we set $f'(x) = 0$ and solve for x.

$0 = \dfrac{x}{\sqrt{400 + x^2}} + \dfrac{x - 25}{\sqrt{900 + (25 - x)^2}}$.

Finding a common denominator:

$0 = \dfrac{x\sqrt{900 + (25 - x)^2} + (x - 25)\sqrt{400 + x^2}}{\sqrt{400 + x^2}\sqrt{900 + (25 - x)^2}}$

261

The numerator must be zero, so we have

$$0 = x\sqrt{900 + (25 - x)^2} + (x - 25)\sqrt{400 + x^2}$$

$$x\sqrt{900 + (25 - x)^2} = (25 - x)\sqrt{400 + x^2}$$

$$\frac{x}{25 - x} = \frac{\sqrt{400 + x^2}}{\sqrt{900 + (25 - x)^2}}$$

(assuming $x \neq 25$).

Squaring both sides, we have

$$\frac{x^2}{625 - 50x + x^2} = \frac{400 + x^2}{900 + (25 - x)^2}$$ and cross-multiplying gives

$$900\,x^2 + 625\,x^2 - 50x^3 + x^4$$

$$= 250,000 - 20,000\,x + 1025\,x^2 - 50x^3 + x^4$$

$$\Rightarrow 500\,x^2 + 20,000\,x - 250,000 = 0$$

$$\Rightarrow x^2 + 40x - 500 = 0$$

$$(x + 50)\,(x - 10) = 0$$

$$x = 10\,, -50\,.$$

We know x is a distance, so it cannot be negative. Hence $x = 10$.

An alternate way to solve the problem is to see that the distance is minimized if the incident angles are equal.

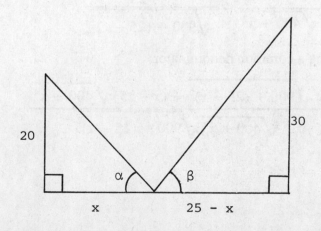

If $\alpha = \beta$, then $\cot \alpha = \cot \beta$

$$\frac{x}{20} = \frac{25 - x}{30}$$

$$30x = 500 - 20x$$

$$50x = 500$$

$$x = 10$$

3.   (A)

The sine of a sum of two angles is given by $\sin (a + b) = \sin a \cos b + \sin b \cos a$.

So $20\sqrt{5} \sin (311t + \theta)$

$\quad = 20\sqrt{5} (\sin (311t) \cos \theta + \sin \theta \cos (311t))$

But $\cos \theta = \dfrac{1}{\sqrt{5}}$ and $\sin \theta = \dfrac{2}{\sqrt{5}}$,   so

$$\quad = 20\sqrt{5} \left(\frac{\sin(311t)}{\sqrt{5}} + \frac{2\cos(311t)}{\sqrt{5}} \right)$$

$$\quad = 20 \sin(311t) + 40 \cos(311t) = I(t) .$$

   (B)

$I(t) = 20\sqrt{5} \sin(311t + \theta)$

$I'(t) = 20\sqrt{5} (311)\cos (311t + \theta) = 0$

$\quad \Rightarrow \cos (311t + \theta) = 0$

$\quad \Rightarrow 311t + \theta = \dfrac{\pi}{2} + \pi k$, for some integer k.

$\quad \Rightarrow t = \dfrac{\dfrac{\pi}{2} - \pi k - \theta}{311}$

$\quad \Rightarrow t = \dfrac{\pi}{622} - \dfrac{\pi k}{311} - \dfrac{\theta}{311}$

263

Now $I''(t) = -20\sqrt{5}\,(311)^2\,\sin(311t + \theta)$

So

$$I''\left(\frac{\pi}{622} - \frac{\pi k}{311} - \frac{\theta}{311}\right)$$

$$= -20\sqrt{5}\,(311)^2\sin\left(\frac{\pi}{2} - \pi k - \theta + \theta\right)$$

$$= -20\sqrt{5}\,(311)^2\sin\left(\frac{\pi}{2} - \pi k\right)$$

$$= -20\sqrt{5}\,(311)^2\left[\sin\left(\frac{\pi}{2}\right)\cos(-\pi k) + \sin(-\pi k)\cos\left(\frac{\pi}{2}\right)\right]$$

$$= -20\sqrt{5}\,(311)^2\,[1 + 0]$$

$$= -20\sqrt{5}\,(311)^2 < 0 ,$$

So we have a maximum by the second derivative test.
The maximum value of $I(t)$ occurs at

$$t = \frac{\pi}{622} - \frac{\pi k}{311} - \frac{\theta}{311}$$

and this maximum value is

$$I\left(\frac{\pi}{622} - \frac{\pi k}{311} - \frac{\theta}{311}\right)$$

$$= 20\sqrt{5}\,\sin\left(\frac{\pi}{2} - \pi k - \theta + \theta\right)$$

$$= 20\sqrt{5}\,\sin\left(\frac{\pi}{2} - \pi k\right)$$

$$= 20\sqrt{5}\left[\sin\left(\frac{\pi}{2}\right)\cos(-\pi k) + \sin(-\pi k)\cos\left(\frac{\pi}{2}\right)\right]$$

$$= 20\sqrt{5} .$$

4. (A)

f is increasing when $f'(x) > 0$. By the chart, we see this occurs between $x = -4$ and $x = 0$, so f is increasing on the interval $[-4, 0]$.

264

f is decreasing when $f'(x) < 0$, which occurs on the intervals $(-\infty, -4]$ and $[0, \infty)$

(B)

Vertical asymptotes occur where the function is undefined. Here $f(x)$ is undefined at $x = -1$, 2 and

$$\lim_{x \to 2^+} f(x) = \infty, \lim_{x \to 2^-} f(x) = -\infty,$$

$$\lim_{x \to -1^+} f(x) = -\infty, \text{ and } \lim_{x \to -1^-} f(x) = \infty,$$

So $x = -1$ and $x = 2$ are vertical asymptotes.

Horizontal asymptotes correspond to $\lim_{x \to \pm\infty} f(x) = y$,

since $\lim_{x \to \pm\infty} f(x) = 0$, we have $y = 0$ as a horizontal asymptote.

(C)

$f(-4) = -\dfrac{2}{3}$ is a local minimum, since $f'(x) < 0$ for $x < -4$ and $f'(x) > 0$ for $x > -4$.

$f(0) = -6$ is a local maximum, since $f'(x) > 0$ for $x < 0$ and $f'(x) < 0$ for $x > 0$.

(D)

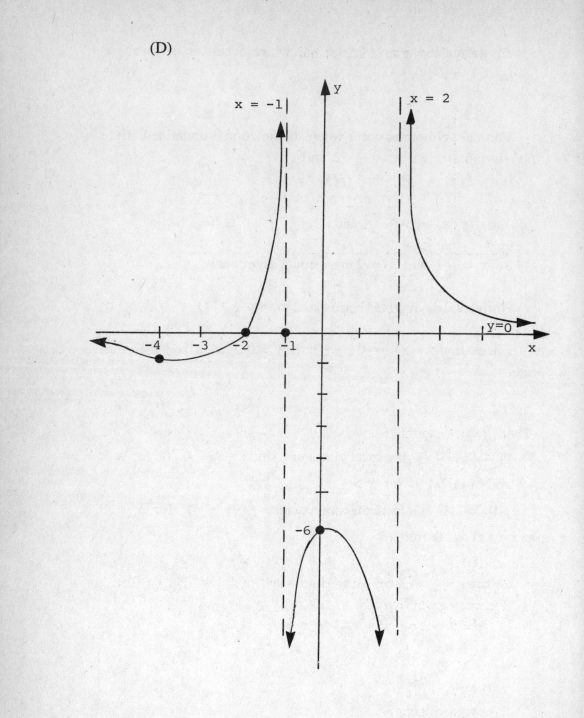

5. (A) Draw graphs of $f(x)$ and $f'(x)$.

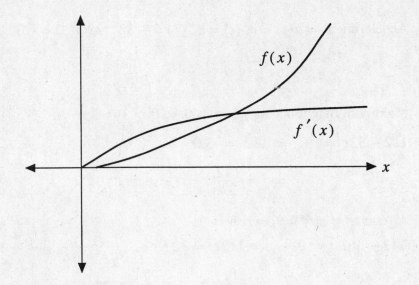

On graph $f'(x)$, by tracing x to $x_0 = 2$, you can find $f'(2)$, $= 1.4$. Then, $c = 1.4$ can make line $y = 1.4x + d$ tangent to $f(x)$ at $x_0 = 2$, if d is found by the following:

$$f(2) = 1.87$$

Hence, $d = y - 1.4x = 1.87 - 1.4 \cdot 2 = -0.93$

(B)

When $c = 1$, we need $f'(x)$, from which $x_0 = 1$ and $f(x_0) = 0.66$. Hence, we can find d by

$$d = y - cx = 0.66 - 1 = -0.34$$

(C)

Use your calculator:

$$fnInt\ (0.667x^\wedge 1.5,\ x,\ 0,\ 3)\ ,$$

which gives 4.16.

6.   (A)
Velocity $= v(t) = s'(t) = 112 - 32t$, so $v(3) = 112 - 96 = 16$.

Acceleration $= a(t) = v'(t) = s''(t) = -32$, so $a(3) = -32$.

(B)
Maximum height occurs when $v(t) = 0$.

$112 - 32t = 0 \Rightarrow 112 = 32t$

$$\Rightarrow t = \frac{112}{32} = \frac{7}{2} \text{ seconds}.$$

Alternately, $s(t) = 0$ when

$112 - 16t^2 = 0 \qquad \Rightarrow 112t = 32t$

$$\Rightarrow \frac{112}{32} = t = \frac{7}{2} \text{ seconds}.$$

The maximum height occurs at the midpoint of the times $t = 0$ and $t = 7$, namely at $t = 7/2$.

(C)
The time of impact occurs when $s(t) = 0$ for the second time.

$16t(7 - t) = 0$

$\Rightarrow t = 0$, $t = 7$.

The velocity at $t = 7$ is $v(7) = s'(7) = 112 - 32(7) = -112$ ft./sec.

(D)
Arc length $= \displaystyle\int_{1}^{10} \sqrt{1 + (y')^2}\ dx$ for the path of the bug.

$y = \frac{1}{3}(x^2 - 2)^{3/2}$

$y' = \frac{1}{3} \cdot \frac{3}{2}(x^2 - 2)^{1/2}\, 2x$

$$= x\sqrt{x^2 - 2}$$

$$(y')^2 = x^2(x^2 - 2)$$

$$= x^4 - 2x^2$$

$$\text{Arclength} = \int_1^{10} \sqrt{1 + x^4 - 2x^2}\ dx$$

$$= \int_1^{10} \sqrt{(x^2 - 1)^2}\ dx$$

$$= \int_1^{10} (x^2 - 1)\ dx$$

$$= \left(\frac{x^3}{3} - x\right)\Bigg|_1^{10}$$

$$= \frac{1000}{3} - 10 - \frac{1}{3} + 1$$

$$= 324 \text{ feet}.$$

The path of the projectile = 392 feet, which is longer than the path of the bug.

THE ADVANCED PLACEMENT EXAMINATION IN

CALCULUS AB

TEST VI

ADVANCED PLACEMENT CALCULUS AB EXAM VI

SECTION I

PART A

Time: 45 minutes
          25 questions

DIRECTIONS: Each of the following problems is followed by five choices. Solve each problem, select the best choice, and blacken the correct space on your answer sheet. Calculators may not be used for this section of the exam.

NOTE:
   Unless otherwise specified, the domain of function f is assumed to be the set of all real numbers x for which $f(x)$ is a real number.

1. If $x < 0$ and $f(x) = |x|$, then $f(f(x))$ is equal to:

(A) x

(B) $-x$

(C) $\dfrac{1}{x}$

(D) $\dfrac{1}{-x}$

(E) Undefined

2. Which one of the following functions satisfies the condition

    that $\int_{-a}^{a} f(x)\ dx = 0$ for any number a?

    (A) $f(x) = x^3 - x^2 + x$

    (B) $f(x) = \dfrac{x^4 + x^3}{x}$

    (C) $f(x) = x^4 - x^2$

    (D) $f(x) = (x + 1)^3 - (3x^2 + 1)$

    (E) None of these

3. Find the area enclosed between the graphs of $x + 2 = y^2$
    and $y = x$.

    (A) $\dfrac{9}{2}$ (D) $\dfrac{26}{6}$

    (B) $\dfrac{7}{2}$ (E) None of these

    (C) $\dfrac{19}{2}$

4. Let $(-2, g(-2))$ be a relative maximum for

    $g(x) = 2x^3 + hx^2 + kx - 6$. Use the fact that

    $\left(-\dfrac{1}{2}, g\left(-\dfrac{1}{2}\right)\right)$ is an inflection point to find the value of

    $(h - k)$.

    (A) 9   (B) -9

274

(C) 15 (D) − 15

(E) 24

5. $\lim\limits_{x \to 0} \dfrac{\sin x}{|x|} =$

(A) − 1 (D) $\dfrac{1}{2}$

(B) 0 (E) None of these

(C) 1

6. Find the instantaneous rate of change of the area of a circle with respect to the circumference C.

(A) C (D) $\dfrac{C}{2\pi}$

(B) $\dfrac{C}{2}$ (E) π

(C) $\dfrac{C}{\pi}$

7. $\lim\limits_{x \to 1} \dfrac{2x - 2}{x^3 + 2x^2 - x - 2} =$

(A) 0 (D) $+\infty$

(B) $\dfrac{1}{3}$ (E) $-\infty$

(C) $\dfrac{2}{3}$

8. Determine which of the following is (are) asymptotes for the graph of $y = \frac{e^x}{x}$:

    I. $x = 0$

    II. $y = 0$

    III. $y = x$

    (A) I only   (D) I and II

    (B) II only   (E) II and III

    (C) III only

9. Let $F(x)$ be an antiderivative of $f(x)$. Suppose $F(x)$ is defined
    by $F(x) = \begin{cases} |x| & \text{if } x < 0 \\ -\sin x & \text{if } x \geq 0 \end{cases}$
    Evaluate $[\, f(b) - f(a)\,]$ for $a = -\frac{\pi}{2}$ and $b = \frac{\pi}{2}$.

    (A) -1 (D) $\frac{\pi + 1}{2}$

    (B) 0 (E) $\frac{\pi + 1}{-2}$

    (C) 1

10. Let $g(x) = f'(x)$ where $f(x) = \cos(\arcsin x)$. Which one of the following statements is <u>FALSE</u> concerning $g(x)$?

(A) The domain of g is $[-1, 1]$.

(B) The range of g is $(-\infty, +\infty)$.

(C) g is a decreasing function.

(D) g is concave down for $x > 0$.

(E) $(0,0)$ is a point of inflection for $y = g(x)$.

11. Let $f(x) = \sqrt{2-x}$. Then $\lim\limits_{x \to 2^-} f'(x) =$

(A) 0   (D) $+\infty$

(B) 1   (E) $-\infty$

(C) -1

12. Let $f(x) = \dfrac{\frac{1}{x} - x}{\frac{1}{x} + x}$. Then $f'(2.5)$ is approximately:

(A) -0.190 (D) 1.005

(B) 0   (E) None of the above

(C) 0.190

277

13. Let $R = \int_{1/\sqrt{2}}^{1} \dfrac{2x}{\sqrt{1-x^4}} \, dx$ and find the interval that contains R.

(A) $(-\infty, .5]$

(D) $(1.5, 2]$

(B) $(.5, 1]$

(E) $(2, +\infty)$

(C) $(1, 1.5]$

14. Suppose a particle moves on a straight line with a position function of $s(t) = 3t^3 - 11t^2 + 8t$. In what interval of time is the particle moving to the left on the line?

(A) $(-\infty, 0)$

(D) $(\dfrac{4}{9}, 2)$

(B) $(0, 1)$

(E) $(2, +\infty)$

(C) $(1, \dfrac{8}{3})$

15. Let $f(x) = (x+2)^3(3-2x)^5(2x-1)^{-3}(3x-4)^{-2}$.
Find $f'(1)$.

(A) -270

(D) 135

(B) -243

(E) None of these

(C) 54

16. $\int (\csc^2 x)\, 2^{\cot x}\, dx =$

(A) $\dfrac{2^{\cot x}}{\cot x\,(\ln 2)} + C$

(B) $\dfrac{2\csc^2 x}{(\ln 2)\cot x} + C$

(C) $\dfrac{-2^{\cot x}}{\ln 2} + C$

(D) $\dfrac{2^{\cot x}\csc^2 x}{\cot x\,(\ln 2)} + C$

(E) $\dfrac{1}{\cot x\,(\ln 2)} + C$

17. Let $f(x) = x^2 + 1$ and $g(x) = \dfrac{1}{x - 2}$. Which one of the following statements is FALSE?

(A) $g(f(x))$ is continuous at $x = 2$.

(B) $f(g(x))$ is continuous at $x = 1$.

(C) $g(f(x))$ has two points of discontinuity.

(D) $\lim\limits_{x \to \infty} f(g(x)) = 1$

(E) $D_x\,[g(f(2))] = -4$

18. Assume $g(x)$ is a continuous function on the reals where $g'(x) > 0$ and $g''(x) > 0$ for $x < a$. Also $g''(x) < 0$ and $g'(x) > 0$ for $x > a$. Further, assume that $g'(a)$ and $g''(a)$ do not exist. Which of the following statements is true about the point $(a, g(a))$?

(A) It is a relative minimum.

(B) It is a relative maximum.

(C) It is a point of inflection.

(D) $y = g(a)$ is an asymptote.

(E) None of these.

19. Use $f(x) = \begin{cases} 2 - x^2 & \text{for } x \geq 0 \\ 2 + x & \text{for } x < 0 \end{cases}$

and find $\lim\limits_{h \to 0} \dfrac{f(x + h) - f(x)}{h}$.

(A) 0   (D) 2

(B) 1   (E) None of these

(C) −1

20. $\displaystyle\lim_{n \to \infty} \left[1 - n\left(\sin \tfrac{1}{n}\right)^2\right] =$

(A) 0

(D) ∞

(B) −1

(E) None of the above

(C) 1

21. Let the velocity of a point moving on a line at time t be defined by $v(t) = 2^t \ln 2 \, \text{cm}/\text{sec}$. How many centimeters did the point travel in the first two seconds?

(A) 3

(D) $4 \ln 2$

(B) 4

(E) $\dfrac{5}{2} \ln 2$

(C) $\dfrac{2}{3} \ln 2$

22. Find the equation of the tangent line to the graph of $y = \dfrac{\ln x}{e^x}$ using $(1, 0)$ as the coordinates of the point of tangency.

(A) $x - ey - 1 = 0$

(B) $x + ey - 1 = 0$

(C) $x - y - 1 = 0$

(D) $x + y - 1 = 0$

(E) None of these

23. Find the average value for $y = \dfrac{e^{\sqrt{x}}}{\sqrt{x}}$ in the interval $[1, 4]$.

(A) 3.114

(B) 34.587

(C) 9.342

(D) 103.760

(E) 0.778

24. $\displaystyle \lim_{x \to a} \frac{\sqrt[3]{x} - \sqrt[3]{a}}{x - a} =$

(A) 0

(B) $2\sqrt[3]{a}$

(C) $\dfrac{3}{2}\sqrt[3]{a^2}$

(D) $\dfrac{\sqrt[3]{a}}{3a}$

(E) None of these

25. $\displaystyle \lim_{x \to \infty} \frac{(\ln x)^2}{x} =$

(A) ∞

(B) 1

(C) $\ln 2$

(D) 2

(E) 0

PART B

Time: 45 minutes

15 questions

DIRECTIONS: Calculators may be used for this section of the test. Each of the following problems is followed by five choices. Solve each problem, select the best choice, and blacken the correct space on your answer sheet.

NOTES:

1. Unless otherwise specified, answers can be given in unsimplified form.

2. The domain of function f is assumed to be the set of all real numbers x for which $f(x)$ is a real number.

26. Let $f(x) = \sin |x|$ and determine which one of the following statements is TRUE.

    (A) $f(x) \geq 0$.

    (B) f is an odd function.

    (C) $\int_{-\frac{\pi}{4}}^{\frac{\pi}{4}} f(x)\, dx = 0$.

    (D) f is symmetric with respect to the line $x = 0$.

    (E) f is differentiable at $x = 0$.

27. Let $h(x) = \dfrac{f(g(x)) - g(f(x))}{f(x)}$ where $f(x) = x - 1$ and $g(x) = x^2$ and x is any real number. What is the range of h?

    (A) All reals   (D) Negative reals

    (B) All reals except 1   (E) None of these

    (C) Positive reals

28. Find the volume of the solid of revolution generated when the region enclosed by the graphs of $x = y^2$ and $x = 2y$ is revolved about the y-axis.

(A) 4.189

(D) 4.114

(B) 8.378

(E) −1.269

(C) 13.404

29. Let $f(x) = \sin^2 x \cos^2 2x$. $\displaystyle\int_0^2 f(x)$ equals

(A) 0.715

(D) 0.015

(B) 1.211

(E) 4.782

(C) 3.121

30. $f(x) = \dfrac{2x}{\sqrt{1 - x^4}}$. The minimum of $f'(x)$.

(A) 1

(D) 2

(B) 0

(E) 3

(C) −1

31. Let $f(x) = -x + x \ln x$ and calculate $D_x\left[f^{-1}(0)\right]$.

(A) 0

(D) e^{-1}

(B) 1

(E) None of these

(C) e

32. Let $y = u^5$, $\dfrac{du}{dx} = 2$, $\dfrac{d^2u}{dx^2} = -3$, and $\dfrac{d^3u}{dx^3} = 5$.

Find the value of $\dfrac{d^3y}{dx^3}$ at $u = 1$.

(A) -20

(D) 145

(B) -35

(E) None of these

(C) 25

33. Let f be differentiable for all reals with critical values at $x = 6$ and $x = -12$. For what values of x will $f'\left(\dfrac{x}{3}\right) = 0$?

(A) 0 and -2

(D) 6 and -12

(B) 2 and -4

(E) 18 and -36

(C) -2 and 4

34. Suppose a particle moves on a straight line with a position fraction of $s(t) = 3t^3 - 11t^2 + 8t$. The highest velocity with which the particle moves in the negative direction is

(A) –5.4

(D) –4

(B) 0

(E) 2

(C) 2.5

35.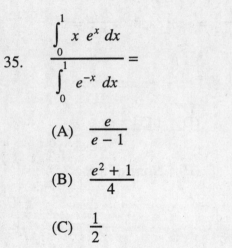

$$\frac{\int_0^1 x\, e^x\, dx}{\int_0^1 e^{-x}\, dx} =$$

(A) $\dfrac{e}{e-1}$

(D) $\dfrac{e^2}{e-1}$

(B) $\dfrac{e^2+1}{4}$

(E) None of these

(C) $\dfrac{1}{2}$

36. Let $f(x) = \ln(\ln x)$ and find the domain of $f(x)$ in interval notation.

(A) $(0, +\infty)$

(D) $[1, +)$

(B) $[0, +)$

(E) None of these

(C) $(1, +)$

37. Let f be a continuous, one-to-one function such that $f(1) = e^{-1}$, $f^{-1}(1) = 0$ and $f'(1) = -2e^{-1}$. Which of the following statements are true?

    I. f is decreasing.

    II. f^{-1} is decreasing and one-to-one.

    III. $D_x \left[f^{-1}(e^{-1}) \right] = -\dfrac{e}{2}$.

(A) I and II   (D) I, II, and III

(B) I and III   (E) None of these

(C) II and III

38. $\displaystyle\int_0^2 x^x$ is

(A) 3.27   (D) 1.98

(B) 2.83   (E) 3.02

(C) 4.21

39. Let $\alpha = \angle BAC$ in $\triangle ABC$ with $\overline{AB} = c$ and $\overline{AC} = b$, where b and c are constants and $c > b$. Side $\overline{BC}$ changes length as the measure of α changes. Find the instantaneous rate of change of the area of $\triangle ABC$ when $\alpha = \pi/3$. Assume the instantaneous rate of change of α is 2.

(A) $cb\sqrt{2}$ (D) $\dfrac{cb}{\sqrt{2}}$

(B) $cb\sqrt{3}$ (E) $\dfrac{cb}{2}$

(C) $\dfrac{cb\sqrt{3}}{2}$

40. Let $f(x) = \dfrac{e^{\sqrt{x}}}{\sqrt{x}}$. $f'(c) = 0$. Then c equals

(A) 3 (D) 2

(B) 0 (E) 7

(C) 1

ADVANCED PLACEMENT CALCULUS AB EXAM VI

SECTION II

Time: 1 hour and 30 minutes
       6 problems

DIRECTIONS: Show all your work. Grading is based on the methods used to solve the problem as well as the accuracy of your final answers. Please make sure all procedures are clearly shown.

NOTES:
1. Unless otherwise specified, answers can be given in unsimplified form.

2. The domain of function f is assumed to be the set of all real numbers x for which $f(x)$ is a real number.

1. Let $f(x) = \dfrac{x^2 - 2x + 1}{2 + x - x^2}$ for x in $(-\infty, +\infty)$.

   (A) Find the critical values of f.

   (B) Sketch the graph of f; label local extrema and asymptotes.

   (C) f has one point of inflection at $(p, f(p))$. Find two consecutive integers n and $n+1$ such that
   $n < p < n + 1$.

2. Population growth in a certain bacteria colony is best described by the equation

$$y = t^2 e^{3t^2} \cdot \sqrt{t}$$

(A) Find the rate of growth at $t = 1$.

(B) Find the lowest rate for $t > 0$.

(C) Find the highest rate for $t > 0$.

3. Let $f(x) = (e^{-\cos x}) \sin x$.

(A) Find $f'(x)$.

(B) Find $f''(x)$.

(C) Use parts (A) and (B), together with symmetry and axes intercepts, to sketch the graph of $y = f(x)$. Make sure your graph depicts the correct concavity.

(D) Evaluate $\displaystyle\int_0^a f(x) \, dx$, where a is the first point of inflection of $f(x)$ in the interval $(0, \pi)$.

4. Find a third degree polynomial function given the following information:

        (i)   The axes intercepts are $(1 , 0)$ and $(0, 12)$.

        (ii)   Relative maximum at $x = -\dfrac{2}{3}$.

        (iii) Point of inflection at $x = \dfrac{5}{3}$.

(A) For what values of x is the function positive?

(B) For what values of x is the derivative of the function positive?

(C) Find the interval(s) in which the function is concave down.

5. Make a rain gutter from a long strip of sheet metal of width w inches using the following prescribed methods. In each case, find the dimension across the top to maximize the amount of rainwater the gutter can handle.

(A) Bend the metal in the middle to form a V-shaped (isosceles $\triangle$) rain gutter. Find the value of x that maximizes the amount of water the gutter can handle by maximizing the cross-sectional area of the gutter.

(B) Bend the metal in two places to form an isosceles trapezoid as follows:

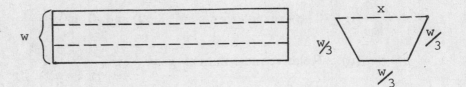

Find the value of x that will maximize the amount of water the gutter can handle by maximizing the cross-sectional area of the gutter.

6.   Let $f(x)$ be continuous on $[-a, a]$ where $a > 0$, $f(a) = 2$ and $f'(a) = 1$.

(A) Find an equation for the tangent line to $y = f(x)$ at the point $(a, f(a))$ and write the equation in slope y-intercept form.

(B) Suppose $f(x)$ is an odd function, i.e. $f(-x) = -f(x)$, and $F(x)$ is an antiderivative for f. Find $F(-a) - F(a)$. Be sure to show work to justify your conclusions.

(C) If the graph of $f(x)$ lies below the tangent line to $y = f(x)$ at $(a, f(a))$ for all x in $[-a, a]$, then find the area between $y = f(x)$ and the tangent line from $x = -a$ to $x = a$ as a function of a.

ADVANCED PLACEMENT
CALCULUS AB
EXAM VI

ANSWER KEY

Section I

1.	B	21.	A
2.	D	22.	A
3.	A	23.	A
4.	C	24.	D
5.	E	25.	E
6.	D	26.	D
7.	B	27.	E
8.	D	28.	C
9.	C	29.	A
10.	A	30.	D
11.	E	31.	B
12.	A	32.	D
13.	C	33.	E
14.	D	34.	A
15.	B	35.	A
16.	C	36.	C
17.	E	37.	D
18.	C	38.	B
19.	E	39.	E
20.	C	40.	C

Section II

See Detailed Explanations of Answers.

ADVANCED PLACEMENT CALCULUS AB EXAM VI

SECTION I

DETAILED EXPLANATIONS OF ANSWERS

1. (B)

    If $x < 0$ and $f(x) = x$ then

    $$f(f(x)) = f(|x|)$$

    $$= f(-x) \text{ since } x < 0$$

    $$= |-x| \text{ because } f(u) = |u|$$

    $$= -x \text{ since } -x > 0.$$

2. (D)

    None of the functions are odd except $f(x) = (x + 1)^3 - (3x^2 + 1)$

    because $f(x) = x^3 + 3x^2 + 3x + 1 - 3x^2 - 1$

    $$= x^3 + 3x.$$

Then $f(-x) = (-x)^3 + 3(-x)$

$$= -x^3 - 3x$$

$$= -f(x).$$

Since $f(x)$ is odd, $\displaystyle\int_{-a}^{a} f(x)\, dx = 0$

3.   (A)

$$\int_{-1}^{2} [y - (y^2 - 2)]\, dy = \left(\frac{y^2}{2} - \frac{y^3}{3} + 2y\right)\Big|_{-1}^{2}$$

$$= \frac{9}{2}.$$

4.   (C)

$$g(x) = 2x^3 + hx^2 + kx - 6$$

so $g'(x) = 6x^2 + 2hx + k$ and

$$g''(x) = 12x + 2h$$

Let $x = -2$ and $g'(-2) = 0$

to get $0 = 24 - 4h + k$

Let $x = -\dfrac{1}{2}$ and $g''(-\dfrac{1}{2}) = 0$

to get $0 = -6 + 2h$ so $3 = h$

Now $k = -12$ and $h - k = 3 - (-12) = 15$

5.   (E)

If $x > 0$, $\displaystyle\lim_{x\to 0} \frac{\sin x}{|x|} = \lim_{x\to 0} \frac{\sin x}{x}$

$$= 1$$

If $x < 0$, $\displaystyle\lim_{x\to 0} \frac{\sin x}{|x|} = \lim_{x\to 0} \frac{\sin x}{-x}$

$$= -1$$

So $\displaystyle\lim_{x\to 0} \frac{\sin x}{|x|}$ does not exist.

6.   (D)

$A = \pi r^2$ and $C = 2\pi r$

$$\Rightarrow \frac{C}{2\pi} = r$$

$$A = \pi\left(\frac{C}{2\pi}\right)^2 = \frac{C^2}{4\pi}$$

so $\displaystyle\frac{dA}{dC} = \frac{2C}{4\pi}$

$$= \frac{C}{2\pi}$$

7.   (B)

$\displaystyle\lim_{x\to 1} \frac{2x-2}{x^3 + 2x^2 - x - 2} = \lim_{x\to 1} \frac{2(x-1)}{(x-1)(x+1)(x+2)}$

$$= \lim_{x\to 1} \frac{2}{(x+1)(x+2)}$$

$$= \frac{1}{3}$$

8.   (D)

$f(x) = \frac{e^x}{x} \Rightarrow f$ is undefined at $x = 0$, so we have a vertical asymptote at $x = 0$.

Also, $\lim\limits_{x \to -\infty} f(x) = 0$, so there is a horizontal asymptote at $y = 0$.

We see that $\lim\limits_{x \to +\infty} f(x) = + \infty$ so there are no other asymptotes.

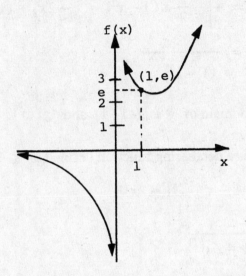

9.   (C)

$$F(x) = \begin{cases} |x| & \text{if } x < 0 \\ - \sin x & \text{if } x \geq 0 \end{cases}$$

$$f(x) = F'(x) = \begin{cases} -1 & \text{if } x < 0 \\ - \cos x & \text{if } x \geq 0 \end{cases}$$

$$f\left(\frac{\pi}{2}\right) - f\left(-\frac{\pi}{2}\right) = - \cos \frac{\pi}{2} - (-1)$$
$$= 1$$

10. (A)

$f(x) = \cos(\arcsin x)$ so

$$f'(x) = -\sin(\arcsin x)\ \frac{1}{\sqrt{1-x^2}} = g(x)$$

$$\frac{-x}{\sqrt{1-x^2}} = g(x)$$

Now $g'(x) = -\left[-\dfrac{1}{2}\,x\,(1-x^2)^{-3/2}(-2x) + (1-x^2)^{-1/2}\right]$

$$= \frac{-1}{(1-x^2)^{3/2}}$$

Note that the domain of g is $(-1, 1)$ and $g'(x) < 0$.

Therefore g is a decreasing function defined on $(-1, 1)$

$g''(x) = \dfrac{3}{2}(1-x^2)^{-5/2}(-2x)$

$$= \frac{-3x}{(1+x^2)^{5/2}}$$

$\Rightarrow g''(0) = 0$, so 0 is an inflection point.

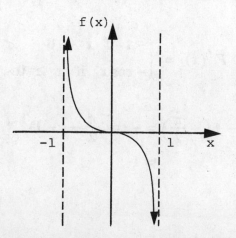

11. (E)

If $f(x) = \sqrt{2-x}$ then $f'(x) = \dfrac{-1}{2\sqrt{2-x}}$.

If $x < 2$ then $0 < 2 - x$ so

$$\lim_{x \to 2^-} f'(x) = -\frac{1}{2} \lim_{x \to 2^-} \frac{1}{\sqrt{2-x}}$$

$$= -\infty$$

12. (A)

$$f(x) = \frac{1-x^2}{1+x^2}$$

$$f'(x) = \frac{\left(1+x^2\right)(-2x) - \left(1-x^2\right)(2x)}{\left(1+x^2\right)^2}$$

$$= \frac{-4x}{\left(1+x^2\right)^2}$$

$$f'(2.5) = \frac{-4(2.5)}{\left(1+2.5^2\right)^2} \approx \frac{-10}{52.563}$$

$$\approx -0.190$$

13.

$$R = \int_{1/\sqrt{2}}^{1} \frac{2x}{\sqrt{1-x^4}} \, dx \qquad \text{Let } u = x^2 \text{ and } du = 2x \, dx$$

$$R = \int_{1/2}^{1} \frac{du}{\sqrt{1-u^2}} \, dx = \arcsin u \, \Big|_{1/2}^{1}$$

$$= \frac{3\pi}{6} - \frac{\pi}{6}$$

$$= \frac{\pi}{3} \approx 1.05 \in (1, 1.5]$$

14.　(D)

The particle is moving to the left when $v(t) < 0$

$$v(t) = s'(t) = 9t^2 - 22t + 8$$

$$= (9t - 4)(t - 2)$$

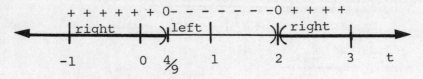

The particle is moving to the left in $\left(\dfrac{4}{9}, 2\right)$

15.　(B)

Let $f(x) = (x+2)^3(3-2x)^5(2x-1)^{-3}(3x-4)^{-2}$

$\ln y = 3\ln(x+2) + 5\ln(3-2x) - 3\ln(2x-1) - 2\ln(3x-4)$

$$\frac{1}{y} y' = \frac{3}{x+2} + \frac{5(-2)}{3-2x} + \frac{-3(2)}{2x-1} + \frac{-2(3)}{3x-4}$$

$$f'(1) = f(1)\left[\frac{3}{3} + \frac{-10}{1} + \frac{-6}{1} + \frac{-6}{-1}\right]$$

$$= (3^3)(1^5)(1)^{-3}(-1)^{-2}(-9)$$

$$= -27(9)$$

$$= -243$$

16. (C)

$$\int (\csc^2 x)\, 2^{\cot x}\, dx \qquad \text{Let } u = \cot x \text{ and } du = -\csc^2 x\, dx$$

$$-\int 2^u\, du = \frac{-2^u}{\ln 2} + C$$

$$= \frac{-2^{\cot x}}{\ln 2} + C$$

17. (E)

Let $f(x) = x^2 + 1$ and $g(x) = \dfrac{1}{x - 2}$.

$$f(g(x)) = f\left(\frac{1}{x-2}\right) = \left(\frac{1}{x-2}\right)^2 + 1,$$

which is continuous at $x = 1$

$$g(f(x)) = g(x^2 + 1) = \frac{1}{x^2 - 1},$$

which is continuous at $x = 2$ and discontinuous at $x = 1$ and -1.

$$\lim_{x \to \infty} f(g(x)) = \lim_{x \to \infty} \left[\left(\frac{1}{x-2}\right)^2 + 1\right]$$

$$= 0^2 + 1$$

$$= 1$$

But $D_x[g(f(x))] = -1(x^2 - 1)^{-2}(2x),$

and $D_x[g(f(2))] = -1(2^2 - 1)^{-2}(2)(2)$

$$= -\frac{4}{9}, \text{ not } -4.$$

18.　　(C)

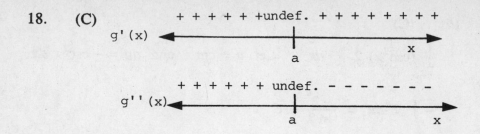

Since $g'(x) > 0$, g is increasing. $g''(x)$ changes sign for values of x on either side of a, therefore $(a, g(a))$ is an inflection point. $g'(a)$ is undefined at $x = a$, but it is continuous at $x = a$, so the graph has a vertical tangent at $x = a$.

$y = \dfrac{1}{3} x^{-2/3}$ is an example of such a function.

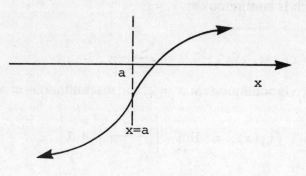

19.　　(E)

$$f(x) = \begin{cases} 2 - x^2 & \text{for } x \geq 0 \\ 2 + x & \text{for } x < 0 \end{cases} \quad \text{and } \lim_{h \to 0} \frac{f(x+h) - f(x)}{h}$$

is the right-hand derivative of $f(x)$ at $x = 0$ which is $D_x(2 - x^2)$ or $-2x$. Thus the right-hand derivative at $x = 0$ is 0. The left-hand derivative at $x = 0$ is $D_x(2 + x) = 1$. Therefore the two-sided limit, which is the derivative, does not exist since the left-hand and right-hand limits are not equal.

20. (C)

$$\lim_{n \to \infty} \left[1 - n \left(\sin \frac{1}{n}\right)^2\right] = 1 - \lim_{n \to \infty} n \left(\sin \frac{1}{n}\right)^2.$$

Let $x = \dfrac{1}{n}$.

Then as $n \to \infty$, we see $x \to 0$.

So $\displaystyle\lim_{n \to \infty} \left[1 - n \left(\sin \frac{1}{n}\right)^2\right]$

$$= 1 - \lim_{x \to 0} \frac{(\sin x)^2}{x}$$

$$= 1 - \lim_{x \to 0} \left(\frac{\sin x}{x}\right)(\sin x)$$

$$= 1 - \left(\lim_{x \to 0} \frac{\sin x}{x}\right)\left(\lim_{x \to 0} \sin x\right)$$

$$= 1 - 1 \cdot 0$$

$$= 1 - 0$$

$$= 1.$$

21. (A)

The position of the point at time t is given by

$$s(t) = \int 2^t \ln 2 \, dt = 2^t + C$$

$$s(0) = 1 + C$$

$$s(2) = 2^2 + C$$

distance $= s(2) - s(0) = 4 + C - 1 - C = 3$

22. (A)

$$\frac{\frac{1}{x}e^x - e^x \ln x}{e^{2x}x} = \frac{\frac{e^x}{x}(1 - x\ln x)}{e^{2x}x}$$

$$= \frac{1 - x\ln x}{xe^x}$$

So when $x = 1$, $y' = \frac{1}{e}$, and the equation of the tangent line

is $y - 0 = \frac{1}{e}(x - 1)$ or $x - ey - 1 = 0$

23. (A)

Find the average value for $y = \dfrac{e^{\sqrt{x}}}{\sqrt{x}}$ in the interval $[1, 4]$

$$\text{Average value} = \frac{1}{4 - 1}\int \frac{e^{\sqrt{x}}}{\sqrt{x}}dx = \frac{2}{3}\int_1^2 e^u \, du$$

$$= \frac{2e}{3}(e - 1)$$

where we let $u = \sqrt{x}$, so $du = \dfrac{dx}{2\sqrt{x}}$

$$= \approx 3.114$$

24. (D)

$$\lim_{x \to a} \frac{\sqrt[3]{x} - \sqrt[3]{a}}{x - a} = f'(a) = D_x\left(\sqrt[3]{x}\right)\Big|_{x=a}$$

$$= \frac{1}{3}a^{-2/3}$$

$$= \frac{\sqrt[3]{a}}{3a}$$

25. (E)

$$\lim_{x \to \infty} \frac{(\ln x)^2}{x} = \frac{\infty}{\infty} \text{ , an indeterminate form.}$$

By L'Hopital's rule, we have $\lim\limits_{x \to \infty} \dfrac{(\ln x)^2}{x} = \lim\limits_{x \to \infty} \dfrac{2(\ln x)x^{-1}}{1}$

$$= \lim_{x \to \infty} \frac{2 \ln x}{x}$$

$$= \frac{\infty}{\infty} \text{ ,}$$

still indeterminate.

Applying L'Hopital's rule again, we have $\lim\limits_{x \to \infty} \dfrac{\frac{2}{x}}{1} = \lim\limits_{x \to \infty} \dfrac{2}{x}$

$$= 0 \text{ .}$$

26. (D)

$f(x) = \sin|x|$

$f(x) < 0$ for x in $\left(\dfrac{\pi}{2}, \pi\right)$ and f is even, not odd.

$$\int_{-\pi/4}^{\pi/4} f(x)\, dx = 2\int_{0}^{\pi/4} f(x)\, dx \neq 0$$

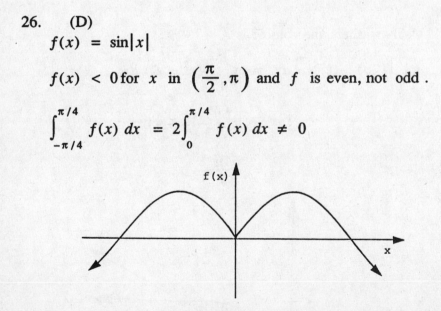

f is symmetric with respect to $x = 0$ since f is even. f is not differentiable at $x = 0$ because the right-hand and left-hand derivatives are $+1$ and -1, respectively.

27. (E)

$$h(x) = \frac{f(g(x)) - g(f(x))}{f(x)}$$

$$= \frac{(x^2 - 1) - (x - 1)^2}{x - 1}$$

$$= \frac{(x + 1)(x - 1) - (x - 1)(x - 1)}{(x - 1)}$$

$$= \frac{(x - 1)[(x + 1) - (x - 1)]}{(x - 1)}$$

$$= [x + 1 - x + 1]$$

$$= 2 .$$

Since h is a constant function, the range of h is $\{ 2 \}$.

28. (C)

Using washers, the volume is

$$\pi \int_0^2 \left[(2y)^2 - (y^2)^2 \right] dy = \pi \int_0^2 (4y^2 - y^4) \, dy$$

$$= \pi \left(\frac{4y^3}{3} - \frac{y^5}{5} \right) \Big|_0^2$$

$$= \frac{64\pi}{15}$$

Using shells, the volume is

$$2\pi \int_0^4 x\left(x^{1/2} - \frac{1}{2}x\right) dx = 2\pi \int_0^4 \left(x^{3/2} - \frac{1}{2}x^2\right) dx$$

$$= \frac{64\pi}{15}$$

Calculator: $64 \times \pi = \div 15 = \approx 13.404$

29.   (A)

This is a direct calculator problem. For example,

fnInt ((sin *x* cos 2*x*)^2, *x*, 0, 2)

can easily give the answer 0.715.

30.   (D)

Draw the graph of $f'(x)$.

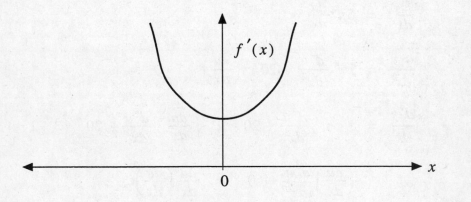

Obviously, a minimum exists at $x = 0$. By tracing on the graph to $x = 0$, you can find

$$f'_{min}(x) = 2.$$

31.   (B)

$f(x) = -x + x\ln x$ and $f(e) = -e + e = 0$ so $e = f^{-1}(0)$

Also $f'(x) = -1 + \frac{x}{x} + \ln x = \ln x$.

$$D_x(f^{-1}(x)) = \frac{1}{f'(f^{-1}(x))}$$

$$D_x(f^{-1}(0)) = \frac{1}{f'(f^{-1}(0))}$$

$$= \frac{1}{f'(e)}$$

$$= \frac{1}{\ln e}$$

$$= 1$$

32.   (D)

$$\frac{dy}{dx} = 5u^4\frac{du}{dx}$$

$$\frac{d^2y}{dx^2} = 5u^4\frac{d^2u}{dx^2} + 20u^3\left(\frac{du}{dx}\right)^2$$

$$\frac{d^3y}{dx^3} = 5 \cdot u^4\frac{d^3u}{dx^3} + 20 \cdot u^3 \cdot \frac{du}{dx} \cdot \frac{d^2u}{dx^2} + 20 \cdot u^3 \cdot 2 \cdot$$

$$\left(\frac{du}{dx}\right)\frac{d^2u}{dx^2} + 60 \cdot u^2 \cdot \frac{du}{dx}\left(\frac{du}{dx}\right)^2$$

$$\frac{d^3y}{dx^3} = 5(1)5 + 20(1)(2)(-3) + 40(1)^3(2)(-3)$$

$$+ 60(1)^2 2(2)^2$$

$$= 145$$

33. (E)

$f'(6) = 0$ and $f'(-12) = 0$

$f'\left(\dfrac{x}{3}\right) = 0$ when $\dfrac{x}{3} = 6$ or $\dfrac{x}{3} = -12$ so

$$x = 18 \quad \text{or} \quad x = -36$$

34. (A)

Draw the graphs of $s(t)$ and $s'(t)$.

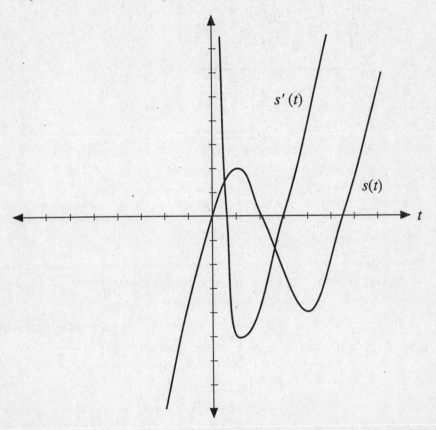

By tracing the most negative $s'(t)$, you can get -5.4 at $t = 1.27$.

35.   (A)

$$\begin{array}{c|c} \text{Let } u = x & du = dx \\ \hline dv = e^x dx & v = e^x \end{array}$$

$$\int_0^1 x\, e^x\, dx = x\, e^x \Big|_0^1 - \int_0^1 e^x\, dx = 1$$

$$\int_0^1 e^{-x}\, dx = -e^{-x}\Big|_0^1$$

$$= -e^{-1} + e^0$$

$$= 1 - e^{-1}$$

$$\therefore \frac{\displaystyle\int_0^1 x\, e^x\, dx}{\displaystyle\int_0^1 e^{-x}\, dx} = \frac{1}{1 - e^{-1}}$$

$$= \frac{e}{e - 1}$$

36.   (C)

Let $f(x) = \ln(\ln x) = \ln u$ where $u > 0$.

If $\ln x > 0$ then $e^{\ln x} > e^0$ because e^x is an increasing function. $\ln x > 0$ means $x > 1$, so the domain is $(1, +\infty)$.

37. (D)

$f^{-1}(1) = 0 \Rightarrow f(0) = 1$

$f(1) = e^{-1} < 1 = f(0) \Rightarrow f(0) > f(1)$,

so f is decreasing. Since f is one-to-one and decreasing, then f^{-1} is also.

$$D_x(f^{-1}(x)) = \frac{1}{f'(f^{-1}(x))} \quad \text{so}$$

$$D_x(f^{-1}(e^{-1})) = \frac{1}{f'(f^{-1}(e^{-1}))}$$

$$= \frac{1}{f'(1)}$$

$$= \frac{1}{-2e^{-1}}$$

$$= -\frac{1}{2}e$$

38. (B)

Use the calculator to solve the problem directly. For example,

$fnInt (x^\wedge x, x, 0, 2)$,

pressing ENTER gives 2.83.

39.   (E)

$$\frac{h}{b} = \sin \alpha \quad \text{so} \quad h = b\sin \alpha$$

$$A = \frac{1}{2} \ (\text{base} \times \text{height})$$

$$= \frac{1}{2} \ cb\sin \alpha$$

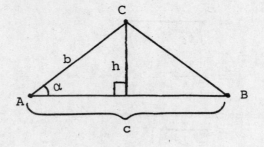

$$\frac{dA}{dt} = \frac{1}{2} \ cb\cos \alpha \left(\frac{d\alpha}{dt}\right)$$

$$= \frac{1}{2} \ cb\cos \left(\frac{\pi}{3}\right)(2)$$

$$= \frac{cb}{2}$$

40.   (C)
   Draw graphs $f(x)$ and $f'(x)$.

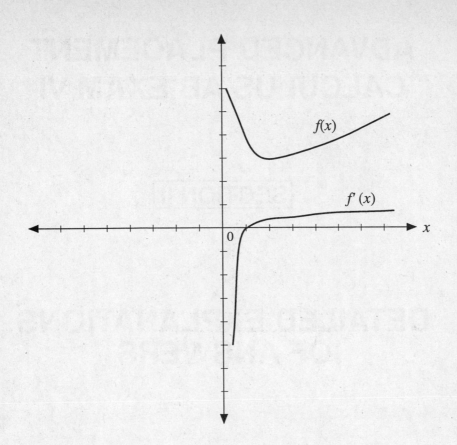

Resetting the viewing window to $[0, 2] \cdot [-1, 1]$, you can easily find that $c = 1$.

ADVANCED PLACEMENT CALCULUS AB EXAM VI

SECTION II

DETAILED EXPLANATIONS OF ANSWERS

1.   (A)

$$f'(x) = \frac{(-x^2 + x + 2)(2x - 2) - (x^2 - 2x + 1)(-2x + 1)}{(x - 2)^2(x + 1)^2}$$

$$= \frac{(x - 1)[-2x^2 + 2x + 4 + 2x^2 - 3x + 1]}{(x - 2)^2(x + 1)^2}$$

$$= \frac{(x - 1)(-x + 5)}{(x - 2)^2(x + 1)^2}$$

The critical values are $x = 5, 2, -1$, and 1, since $f'(x)$ at these points is either zero or undefined.

(B)

$$f(x) = \frac{(x-1)^2}{(2-x)(x+1)}$$

```
+ + + + + + + + + + + +0 - - - -   (2 - x)
    - - - - 0+ + +   + + + + + + + + +(x + 1)
  ←——————————————————————————————————→
      -1   0    1    2                x
```

$$f'(x) = \frac{(x-1)(-x+5)}{(x-2)^2(x+1)^2}$$

```
(x-1)- - - - - - - - - -0 + + + + + + + + + + + + + + +
( x+5)+ + + + + + + + + + + + + + + + + + + + 0 - - -
  ←— decreasing ——— X increasing ——— X decreasing →
                  |      |      |      |      |      x
       0    1    2    3    4    5
```

y = f(x)

(C)

$$f''(x) = \frac{\left[(x-2)^2(x+1)^2\right]D_x\left[(x-1)(-x+5)\right]}{(x-2)^4}$$

$$\frac{-\left[(x-1)(-x+5)\right]D_x\left[(x-2)^2(x+1)^2\right]}{(x+1)^4}$$

$$f''(x) = \frac{2\left[(x-2)(x+1)(-x+3) + (x-1)(x-5)(2x-1)\right]}{(x-2)^3(x+1)^3}$$

$$f''(x) = \frac{2(x^3 - 9x^2 + 15x - 11)}{(x-2)^3(x+1)^3}$$

Using synthetic division,

	1	−9	15	−11
1	1	−8	7	−4
2	1	−7	1	−9
5	1	−4	−5	−36
7	1	−2	1	−4
8	1	−1	7	45
9	1	0	15	+
10	1	1	25	+

x

The second derivative is zero between 7 and 8 so there is a point of inflection between $n = 7$ and $n + 1 = 8$.

The chart shows that

$$f''(7) = \frac{2(7^3 - 9(7^2) + 15(7) - 11)}{(7 - 2)^3(7 + 1)^3}$$

$$= \frac{2(-4)}{5^3 \cdot 8^3} < 0$$

and $f''(8) = \dfrac{2(8^3 - 9(8^2) + 15(8) - 11)}{(8 - 2)^3(8 + 1)^3}$

$$= \frac{2(45)}{6^3 \cdot 9^3} > 0,$$

so $f''(x) = 0$ between $x = 7$ and $x = 8$, indicating an inflection point.

2.   (A)
   Use the calculator directly,

$$\text{der } 1 \; (t^\wedge 2 \; e^\wedge (3t^\wedge 2 + \sqrt{t}), x, 1)$$

gives 464.1.

   (B)
Draw the graph of y.

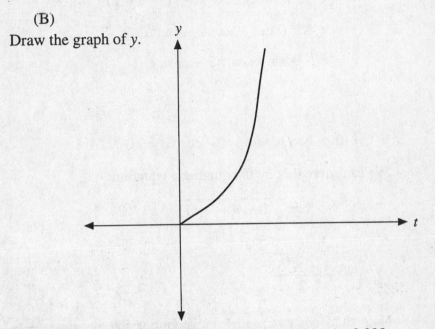

   The lowest is obviously at $t = 0$, when $t = 0.001$, $y' = 0.002$.
Thus, it can be seen from the graph as $t \to 0$, $y' \to 0$.
   So, the lowest rate is 0.

   (C)
Obviously, as $t \to \infty$, $y' \to \infty$.

3.   (A)

$$f'(x) = (e^{-\cos x})(\cos x) + \sin x (e^{-\cos x}) \sin x$$

$$f'(x) = e^{-\cos x}[\cos x + \sin^2 x]$$

$$= e^{-\cos x} [\cos x + (1 - \cos^2 x)]$$

(B)

$$f''(x) = e^{-\cos x}[-\sin x + 2\sin x \cos x + \sin x \cos x + \sin^3 x]$$

$$= e^{-\cos x}[-\sin x + 3\sin x \cos x + \sin^3 x]$$

$$= e^{-\cos x}(\sin x)(3\cos x + \sin^2 x - 1)$$

$$= e^{-\cos x}(\sin x)(3\cos x - (1 - \sin^2 x))$$

$$= e^{-\cos x}(\sin x)(3\cos x - \cos^2 x)$$

$$= e^{-\cos x}\sin x \cos x(3 - \cos x)$$

(C)

$y' = 0$ when $\cos x + 1 - \cos^2 x = 0$.

We can solve this by the quadratic equation:

$$\cos x = -\frac{1 \pm \sqrt{1 - 4(-1)(1)}}{2(-1)}$$

$$= \frac{1 - \sqrt{5}}{2}.$$

Note that $\cos x = \dfrac{1 + \sqrt{5}}{2}$ is not possible because $\dfrac{1 + \sqrt{5}}{2}$ > 1 and $\cos x$ is always ≤ 1.

So $x = \arccos\left(\dfrac{1 - \sqrt{5}}{2}\right) + 2k\pi \leftarrow$ Location of relative extrema for any integer k.

Let $y'' = 0$ to find points of inflection.

$y'' = 0$ when $\sin x = 0$ or $\cos x = 0$ ($3 - \cos x = 0$ is impossible, as is $e^{-\cos x} = 0$).

$\Rightarrow x = k\pi$ or $x = \dfrac{\pi}{2} + k\pi$ are where the points of inflection are found, for any integer k.

318

The function crosses the x-axis when $f(x) = e^{-\cos x}(\sin x) = 0$ i.e. when $\sin x = 0$, namely $x = k\pi$, k = any integer.

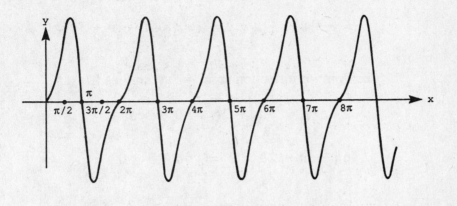

(D) $\displaystyle\int_0^{\pi/2} e^{-\cos x} \sin x \, dx = \int_{-1}^{0} e^u \, du$ where $u = -\cos x$, $du = \sin x \; dx$

$$= e^u \Big|_{-1}^{0}$$

$$= e^0 - e^{-1}$$

$$= 1 - \frac{1}{e}$$

4. (A)

Start with a general third degree polynomial of the form

$$P(x) = ax^3 + bx^2 + cx + d$$

i. The intercepts are $(1, 0)$ and $(0, 12)$.

Let $x = 0$ to get $P(0) = 12 = d$

So $P(x) = ax^3 + bx^2 + cx + 12$

Let $x = 1$ to get $P(1) = 0 = a + b + c + 12$

ii. Relative maximum at $x = -\dfrac{2}{3}$ implies

$$P'(x) = 3ax^2 + 2bx + c = 0 \quad \text{when} \quad x = -\dfrac{2}{3}$$

$$3\left(\dfrac{4}{9}\right)a - \dfrac{4}{3}b + c = 0 \implies 4a - 4b + 3c = 0$$

iii. Point of inflection at $x = \dfrac{5}{3}$ implies

$$P''(x) = 6ax + 2b = 0 \quad \text{when} \quad x = \dfrac{5}{3}$$

$$6\left(\dfrac{5}{3}\right)a + 2b = 0 \implies 5a + b = 0$$

Now solve the equations above simultaneously.

$a + b + c = -12$

$4a - 4b + 3c = 0$

$5a + b = 0$

$\implies 5a = -b$

$a - 5a + c = -12 \implies -4a + c = -12$

$4a + 4(5a) + 3c = 0 \implies 24a + 3c = 0 \implies 8a + c = 0$

$12a = 12 \ , \ a = 1 \ , \ b = -5 \ , \ c = -12 + 4a = -8 \ .$

The simultaneous solution is $a = 1 \, , \ b = -5 \, , \ c = -8$ so

$$P(x) = x^3 - 5x^2 - 8x + 12 = (x - 1)(x - 6)(x + 2)$$

(A)

```
- - 0  + + + + + + + + + + + + + + + + + + + + + +   (x+2)
- - - - - - - - - - - - - - - - - - - - - -0+ + + + +  (x-6)
- - - - - - - - - 0 + + + + + + + + + + + + + + + + +  (x-1)
    ✕   +   +   ✕   +   +   +   +   ✕
   -2  -1   0   1   2   3   4   5   6              x
```

The polynomial is positive for $-2 < x < 1$ or $x > 6$.

(B)

$$P'(x) = 3x^2 - 10x - 8 = (3x + 2)(x - 4)$$

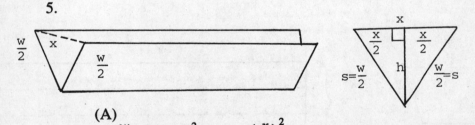

The derivative is positive for $x < -\dfrac{2}{3}$ or $x > 4$.

(C)

$P''(x) = 6x - 10 = 2(3x - 5)$. When $P''(x) < 0$, the function is concave down.

The function is concave down for

$$x < \frac{5}{3}.$$

5.

(A)

If $s = \dfrac{w}{2}$ then $h^2 = s^2 - \left(\dfrac{x}{2}\right)^2$ for $x \in [0, 2s]$, by the Pythagorean theorem. So the cross-sectional area is

$$A = \frac{1}{2}bh = \frac{1}{2}x\sqrt{s^2 - \frac{x^2}{4}}, \quad \text{so } A^2 = \frac{x^2 s^2}{4} - \frac{x^4}{16},$$

which is an easier function to work with.

The maximum value of a continuous function in a closed interval occurs at the end points or at a critical value. So evaluate the function at 0, $2s$, and any critical values.

To find the critical values, we differentiate A^2 with respect to x:

$$2AA' = \frac{s^2}{4} 2x - \frac{4x^3}{16} = \frac{x}{4}(2s^2 - x^2)$$

Set $A' = 0$ to find critical value(s). If $A' = 0$ then $x = 0$ or $2s^2 - x^2 = 0$. $x = 0$ makes no constructive sense, so $x = s\sqrt{2}$.

Evaluate A at the endpoints and critical point to get:

x	A
0	0
$2s$	0
$s\sqrt{2}$	$\dfrac{s^2}{2}$

The maximum amount of water is handled when

$$x = s\sqrt{2} = \frac{w\sqrt{2}}{2}$$

(B)

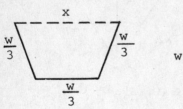

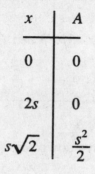

If $a = \dfrac{w}{3}$ then $h^2 = a^2 - y$ for $y \in [0, a]$, by the Pythagorean theorem.

$$A = \frac{1}{2}h(2y + 2a) = \frac{2}{2}\sqrt{a^2 - y^2}(y + a)$$

$$= (y + a)\sqrt{a^2 - y^2}.$$

322

Taking the derivative of A with respect to y, we have

$$A' = (y + a)\frac{1}{2}(a^2 - y^2)^{-1/2}(-2y) + \sqrt{a^2 + y^2}$$

$$A' = (a^2 - y^2)^{-1/2}\left[-y^2 - ay + a^2 - y^2\right]$$

$$= \frac{(-2y^2 - ay + a^2)}{\sqrt{a^2 \quad y^2}}$$

The maximum value of a continuous function in a closed interval occurs at a critical value or its endpoints. So evaluate the function at 0 and a and any critical values. We find critical values as follows:

If $A' = 0$ then $-2y^2 - ay + a^2 = 0$

$$(-2y + a)(y + a) = 0$$

$$\Rightarrow y = \frac{a}{2} \text{ or } y = -a \ .$$

Since $a > 0$, $y = -a$ means $y < 0$ which makes no sense.

So $y = \frac{a}{2}$ is a critical value. We then have:

y	A
0	a^2
a	0
$\frac{a}{2}$	$\frac{3a^2\sqrt{3}}{4} > a^2$

So the maximum amount of water is handled when $y = \frac{a}{2}$, which means $x = a + 2y = a + a = 2a.$

6.   (A)

$$y - y_1 = m(x - x_1) \quad \text{so} \quad y - 2 = 1(x - a)$$

$$y = x - a + 2$$

(B)

$$F(x) = \int f(x)\, dx$$

$$-F(x) = -\int f(x)\, dx$$

$$= \int f(-x)\, dx$$

$$= -\int f(u)\, du$$

$$= -F(u) + C$$

where $u = -x$ and $du = -dx$

$$-F(x) = -F(u) + C = -F(-x) + C \Rightarrow F(x) = F(-x) - C$$

Let $x = 0$ to get $F(0) = F(0) - C$ so $C = 0$.

So $F(x) = F(-x)$ and $F(-a) - F(a) = F(a) - F(a) = 0$

(C)

$$A = \int_{-a}^{a} [(x - a + 2) - f(x)]\, dx = \left[\frac{x^2}{2} + (2 - a) - F(x) \right]\Big|_{-a}^{a}$$

$$A = \left[\frac{a^2}{2} + (2 - a)a - F(a) \right] - \left[\frac{a^2}{2} + (2 - a)(-a) - F(a) \right]$$

$$A = 2a(2 - a) - F(a) + F(-a) = 2a(2 - a)$$

$$\text{since} \quad F(-a) = F(a)$$

THE ADVANCED PLACEMENT EXAMINATION IN

CALCULUS AB

ANSWER SHEETS

ADVANCED PLACEMENT
EXAMINATION
CALCULUS AB
ANSWER SHEET
TEST 1

1. Ⓐ Ⓑ Ⓒ Ⓓ Ⓔ
2. Ⓐ Ⓑ Ⓒ Ⓓ Ⓔ
3. Ⓐ Ⓑ Ⓒ Ⓓ Ⓔ
4. Ⓐ Ⓑ Ⓒ Ⓓ Ⓔ
5. Ⓐ Ⓑ Ⓒ Ⓓ Ⓔ
6. Ⓐ Ⓑ Ⓒ Ⓓ Ⓔ
7. Ⓐ Ⓑ Ⓒ Ⓓ Ⓔ
8. Ⓐ Ⓑ Ⓒ Ⓓ Ⓔ
9. Ⓐ Ⓑ Ⓒ Ⓓ Ⓔ
10. Ⓐ Ⓑ Ⓒ Ⓓ Ⓔ
11. Ⓐ Ⓑ Ⓒ Ⓓ Ⓔ
12. Ⓐ Ⓑ Ⓒ Ⓓ Ⓔ
13. Ⓐ Ⓑ Ⓒ Ⓓ Ⓔ
14. Ⓐ Ⓑ Ⓒ Ⓓ Ⓔ
15. Ⓐ Ⓑ Ⓒ Ⓓ Ⓔ
16. Ⓐ Ⓑ Ⓒ Ⓓ Ⓔ
17. Ⓐ Ⓑ Ⓒ Ⓓ Ⓔ
18. Ⓐ Ⓑ Ⓒ Ⓓ Ⓔ
19. Ⓐ Ⓑ Ⓒ Ⓓ Ⓔ
20. Ⓐ Ⓑ Ⓒ Ⓓ Ⓔ

21. Ⓐ Ⓑ Ⓒ Ⓓ Ⓔ
22. Ⓐ Ⓑ Ⓒ Ⓓ Ⓔ
23. Ⓐ Ⓑ Ⓒ Ⓓ Ⓔ
24. Ⓐ Ⓑ Ⓒ Ⓓ Ⓔ
25. Ⓐ Ⓑ Ⓒ Ⓓ Ⓔ
26. Ⓐ Ⓑ Ⓒ Ⓓ Ⓔ
27. Ⓐ Ⓑ Ⓒ Ⓓ Ⓔ
28. Ⓐ Ⓑ Ⓒ Ⓓ Ⓔ
29. Ⓐ Ⓑ Ⓒ Ⓓ Ⓔ
30. Ⓐ Ⓑ Ⓒ Ⓓ Ⓔ
31. Ⓐ Ⓑ Ⓒ Ⓓ Ⓔ
32. Ⓐ Ⓑ Ⓒ Ⓓ Ⓔ
33. Ⓐ Ⓑ Ⓒ Ⓓ Ⓔ
34. Ⓐ Ⓑ Ⓒ Ⓓ Ⓔ
35. Ⓐ Ⓑ Ⓒ Ⓓ Ⓔ
36. Ⓐ Ⓑ Ⓒ Ⓓ Ⓔ
37. Ⓐ Ⓑ Ⓒ Ⓓ Ⓔ
38. Ⓐ Ⓑ Ⓒ Ⓓ Ⓔ
39. Ⓐ Ⓑ Ⓒ Ⓓ Ⓔ
40. Ⓐ Ⓑ Ⓒ Ⓓ Ⓔ

ADVANCED PLACEMENT EXAMINATION

CALCULUS AB

ANSWER SHEET
TEST 2

1. Ⓐ Ⓑ Ⓒ Ⓓ Ⓔ
2. Ⓐ Ⓑ Ⓒ Ⓓ Ⓔ
3. Ⓐ Ⓑ Ⓒ Ⓓ Ⓔ
4. Ⓐ Ⓑ Ⓒ Ⓓ Ⓔ
5. Ⓐ Ⓑ Ⓒ Ⓓ Ⓔ
6. Ⓐ Ⓑ Ⓒ Ⓓ Ⓔ
7. Ⓐ Ⓑ Ⓒ Ⓓ Ⓔ
8. Ⓐ Ⓑ Ⓒ Ⓓ Ⓔ
9. Ⓐ Ⓑ Ⓒ Ⓓ Ⓔ
10. Ⓐ Ⓑ Ⓒ Ⓓ Ⓔ
11. Ⓐ Ⓑ Ⓒ Ⓓ Ⓔ
12. Ⓐ Ⓑ Ⓒ Ⓓ Ⓔ
13. Ⓐ Ⓑ Ⓒ Ⓓ Ⓔ
14. Ⓐ Ⓑ Ⓒ Ⓓ Ⓔ
15. Ⓐ Ⓑ Ⓒ Ⓓ Ⓔ
16. Ⓐ Ⓑ Ⓒ Ⓓ Ⓔ
17. Ⓐ Ⓑ Ⓒ Ⓓ Ⓔ
18. Ⓐ Ⓑ Ⓒ Ⓓ Ⓔ
19. Ⓐ Ⓑ Ⓒ Ⓓ Ⓔ
20. Ⓐ Ⓑ Ⓒ Ⓓ Ⓔ

21. Ⓐ Ⓑ Ⓒ Ⓓ Ⓔ
22. Ⓐ Ⓑ Ⓒ Ⓓ Ⓔ
23. Ⓐ Ⓑ Ⓒ Ⓓ Ⓔ
24. Ⓐ Ⓑ Ⓒ Ⓓ Ⓔ
25. Ⓐ Ⓑ Ⓒ Ⓓ Ⓔ
26. Ⓐ Ⓑ Ⓒ Ⓓ Ⓔ
27. Ⓐ Ⓑ Ⓒ Ⓓ Ⓔ
28. Ⓐ Ⓑ Ⓒ Ⓓ Ⓔ
29. Ⓐ Ⓑ Ⓒ Ⓓ Ⓔ
30. Ⓐ Ⓑ Ⓒ Ⓓ Ⓔ
31. Ⓐ Ⓑ Ⓒ Ⓓ Ⓔ
32. Ⓐ Ⓑ Ⓒ Ⓓ Ⓔ
33. Ⓐ Ⓑ Ⓒ Ⓓ Ⓔ
34. Ⓐ Ⓑ Ⓒ Ⓓ Ⓔ
35. Ⓐ Ⓑ Ⓒ Ⓓ Ⓔ
36. Ⓐ Ⓑ Ⓒ Ⓓ Ⓔ
37. Ⓐ Ⓑ Ⓒ Ⓓ Ⓔ
38. Ⓐ Ⓑ Ⓒ Ⓓ Ⓔ
39. Ⓐ Ⓑ Ⓒ Ⓓ Ⓔ
40. Ⓐ Ⓑ Ⓒ Ⓓ Ⓔ

ADVANCED PLACEMENT EXAMINATION

CALCULUS AB

ANSWER SHEET
TEST 3

1. Ⓐ Ⓑ Ⓒ Ⓓ Ⓔ
2. Ⓐ Ⓑ Ⓒ Ⓓ Ⓔ
3. Ⓐ Ⓑ Ⓒ Ⓓ Ⓔ
4. Ⓐ Ⓑ Ⓒ Ⓓ Ⓔ
5. Ⓐ Ⓑ Ⓒ Ⓓ Ⓔ
6. Ⓐ Ⓑ Ⓒ Ⓓ Ⓔ
7. Ⓐ Ⓑ Ⓒ Ⓓ Ⓔ
8. Ⓐ Ⓑ Ⓒ Ⓓ Ⓔ
9. Ⓐ Ⓑ Ⓒ Ⓓ Ⓔ
10. Ⓐ Ⓑ Ⓒ Ⓓ Ⓔ
11. Ⓐ Ⓑ Ⓒ Ⓓ Ⓔ
12. Ⓐ Ⓑ Ⓒ Ⓓ Ⓔ
13. Ⓐ Ⓑ Ⓒ Ⓓ Ⓔ
14. Ⓐ Ⓑ Ⓒ Ⓓ Ⓔ
15. Ⓐ Ⓑ Ⓒ Ⓓ Ⓔ
16. Ⓐ Ⓑ Ⓒ Ⓓ Ⓔ
17. Ⓐ Ⓑ Ⓒ Ⓓ Ⓔ
18. Ⓐ Ⓑ Ⓒ Ⓓ Ⓔ
19. Ⓐ Ⓑ Ⓒ Ⓓ Ⓔ
20. Ⓐ Ⓑ Ⓒ Ⓓ Ⓔ

21. Ⓐ Ⓑ Ⓒ Ⓓ Ⓔ
22. Ⓐ Ⓑ Ⓒ Ⓓ Ⓔ
23. Ⓐ Ⓑ Ⓒ Ⓓ Ⓔ
24. Ⓐ Ⓑ Ⓒ Ⓓ Ⓔ
25. Ⓐ Ⓑ Ⓒ Ⓓ Ⓔ
26. Ⓐ Ⓑ Ⓒ Ⓓ Ⓔ
27. Ⓐ Ⓑ Ⓒ Ⓓ Ⓔ
28. Ⓐ Ⓑ Ⓒ Ⓓ Ⓔ
29. Ⓐ Ⓑ Ⓒ Ⓓ Ⓔ
30. Ⓐ Ⓑ Ⓒ Ⓓ Ⓔ
31. Ⓐ Ⓑ Ⓒ Ⓓ Ⓔ
32. Ⓐ Ⓑ Ⓒ Ⓓ Ⓔ
33. Ⓐ Ⓑ Ⓒ Ⓓ Ⓔ
34. Ⓐ Ⓑ Ⓒ Ⓓ Ⓔ
35. Ⓐ Ⓑ Ⓒ Ⓓ Ⓔ
36. Ⓐ Ⓑ Ⓒ Ⓓ Ⓔ
37. Ⓐ Ⓑ Ⓒ Ⓓ Ⓔ
38. Ⓐ Ⓑ Ⓒ Ⓓ Ⓔ
39. Ⓐ Ⓑ Ⓒ Ⓓ Ⓔ
40. Ⓐ Ⓑ Ⓒ Ⓓ Ⓔ

ADVANCED PLACEMENT EXAMINATION

CALCULUS AB

ANSWER SHEET
TEST 4

1. (A) (B) (C) (D) (E)   21. (A) (B) (C) (D) (E)
2. (A) (B) (C) (D) (E)   22. (A) (B) (C) (D) (E)
3. (A) (B) (C) (D) (E)   23. (A) (B) (C) (D) (E)
4. (A) (B) (C) (D) (E)   24. (A) (B) (C) (D) (E)
5. (A) (B) (C) (D) (E)   25. (A) (B) (C) (D) (E)
6. (A) (B) (C) (D) (E)   26. (A) (B) (C) (D) (E)
7. (A) (B) (C) (D) (E)   27. (A) (B) (C) (D) (E)
8. (A) (B) (C) (D) (E)   28. (A) (B) (C) (D) (E)
9. (A) (B) (C) (D) (E)   29. (A) (B) (C) (D) (E)
10. (A) (B) (C) (D) (E)   30. (A) (B) (C) (D) (E)
11. (A) (B) (C) (D) (E)   31. (A) (B) (C) (D) (E)
12. (A) (B) (C) (D) (E)   32. (A) (B) (C) (D) (E)
13. (A) (B) (C) (D) (E)   33. (A) (B) (C) (D) (E)
14. (A) (B) (C) (D) (E)   34. (A) (B) (C) (D) (E)
15. (A) (B) (C) (D) (E)   35. (A) (B) (C) (D) (E)
16. (A) (B) (C) (D) (E)   36. (A) (B) (C) (D) (E)
17. (A) (B) (C) (D) (E)   37. (A) (B) (C) (D) (E)
18. (A) (B) (C) (D) (E)   38. (A) (B) (C) (D) (E)
19. (A) (B) (C) (D) (E)   39. (A) (B) (C) (D) (E)
20. (A) (B) (C) (D) (E)   40. (A) (B) (C) (D) (E)

ADVANCED PLACEMENT EXAMINATION

CALCULUS AB

ANSWER SHEET
TEST 5

1. Ⓐ Ⓑ Ⓒ Ⓓ Ⓔ
2. Ⓐ Ⓑ Ⓒ Ⓓ Ⓔ
3. Ⓐ Ⓑ Ⓒ Ⓓ Ⓔ
4. Ⓐ Ⓑ Ⓒ Ⓓ Ⓔ
5. Ⓐ Ⓑ Ⓒ Ⓓ Ⓔ
6. Ⓐ Ⓑ Ⓒ Ⓓ Ⓔ
7. Ⓐ Ⓑ Ⓒ Ⓓ Ⓔ
8. Ⓐ Ⓑ Ⓒ Ⓓ Ⓔ
9. Ⓐ Ⓑ Ⓒ Ⓓ Ⓔ
10. Ⓐ Ⓑ Ⓒ Ⓓ Ⓔ
11. Ⓐ Ⓑ Ⓒ Ⓓ Ⓔ
12. Ⓐ Ⓑ Ⓒ Ⓓ Ⓔ
13. Ⓐ Ⓑ Ⓒ Ⓓ Ⓔ
14. Ⓐ Ⓑ Ⓒ Ⓓ Ⓔ
15. Ⓐ Ⓑ Ⓒ Ⓓ Ⓔ
16. Ⓐ Ⓑ Ⓒ Ⓓ Ⓔ
17. Ⓐ Ⓑ Ⓒ Ⓓ Ⓔ
18. Ⓐ Ⓑ Ⓒ Ⓓ Ⓔ
19. Ⓐ Ⓑ Ⓒ Ⓓ Ⓔ
20. Ⓐ Ⓑ Ⓒ Ⓓ Ⓔ

21. Ⓐ Ⓑ Ⓒ Ⓓ Ⓔ
22. Ⓐ Ⓑ Ⓒ Ⓓ Ⓔ
23. Ⓐ Ⓑ Ⓒ Ⓓ Ⓔ
24. Ⓐ Ⓑ Ⓒ Ⓓ Ⓔ
25. Ⓐ Ⓑ Ⓒ Ⓓ Ⓔ
26. Ⓐ Ⓑ Ⓒ Ⓓ Ⓔ
27. Ⓐ Ⓑ Ⓒ Ⓓ Ⓔ
28. Ⓐ Ⓑ Ⓒ Ⓓ Ⓔ
29. Ⓐ Ⓑ Ⓒ Ⓓ Ⓔ
30. Ⓐ Ⓑ Ⓒ Ⓓ Ⓔ
31. Ⓐ Ⓑ Ⓒ Ⓓ Ⓔ
32. Ⓐ Ⓑ Ⓒ Ⓓ Ⓔ
33. Ⓐ Ⓑ Ⓒ Ⓓ Ⓔ
34. Ⓐ Ⓑ Ⓒ Ⓓ Ⓔ
35. Ⓐ Ⓑ Ⓒ Ⓓ Ⓔ
36. Ⓐ Ⓑ Ⓒ Ⓓ Ⓔ
37. Ⓐ Ⓑ Ⓒ Ⓓ Ⓔ
38. Ⓐ Ⓑ Ⓒ Ⓓ Ⓔ
39. Ⓐ Ⓑ Ⓒ Ⓓ Ⓔ
40. Ⓐ Ⓑ Ⓒ Ⓓ Ⓔ

ADVANCED PLACEMENT
EXAMINATION
CALCULUS AB
ANSWER SHEET
TEST 6

1. Ⓐ Ⓑ Ⓒ Ⓓ Ⓔ   21. Ⓐ Ⓑ Ⓒ Ⓓ Ⓔ
2. Ⓐ Ⓑ Ⓒ Ⓓ Ⓔ   22. Ⓐ Ⓑ Ⓒ Ⓓ Ⓔ
3. Ⓐ Ⓑ Ⓒ Ⓓ Ⓔ   23. Ⓐ Ⓑ Ⓒ Ⓓ Ⓔ
4. Ⓐ Ⓑ Ⓒ Ⓓ Ⓔ   24. Ⓐ Ⓑ Ⓒ Ⓓ Ⓔ
5. Ⓐ Ⓑ Ⓒ Ⓓ Ⓔ   25. Ⓐ Ⓑ Ⓒ Ⓓ Ⓔ
6. Ⓐ Ⓑ Ⓒ Ⓓ Ⓔ   26. Ⓐ Ⓑ Ⓒ Ⓓ Ⓔ
7. Ⓐ Ⓑ Ⓒ Ⓓ Ⓔ   27. Ⓐ Ⓑ Ⓒ Ⓓ Ⓔ
8. Ⓐ Ⓑ Ⓒ Ⓓ Ⓔ   28. Ⓐ Ⓑ Ⓒ Ⓓ Ⓔ
9. Ⓐ Ⓑ Ⓒ Ⓓ Ⓔ   29. Ⓐ Ⓑ Ⓒ Ⓓ Ⓔ
10. Ⓐ Ⓑ Ⓒ Ⓓ Ⓔ   30. Ⓐ Ⓑ Ⓒ Ⓓ Ⓔ
11. Ⓐ Ⓑ Ⓒ Ⓓ Ⓔ   31. Ⓐ Ⓑ Ⓒ Ⓓ Ⓔ
12. Ⓐ Ⓑ Ⓒ Ⓓ Ⓔ   32. Ⓐ Ⓑ Ⓒ Ⓓ Ⓔ
13. Ⓐ Ⓑ Ⓒ Ⓓ Ⓔ   33. Ⓐ Ⓑ Ⓒ Ⓓ Ⓔ
14. Ⓐ Ⓑ Ⓒ Ⓓ Ⓔ   34. Ⓐ Ⓑ Ⓒ Ⓓ Ⓔ
15. Ⓐ Ⓑ Ⓒ Ⓓ Ⓔ   35. Ⓐ Ⓑ Ⓒ Ⓓ Ⓔ
16. Ⓐ Ⓑ Ⓒ Ⓓ Ⓔ   36. Ⓐ Ⓑ Ⓒ Ⓓ Ⓔ
17. Ⓐ Ⓑ Ⓒ Ⓓ Ⓔ   37. Ⓐ Ⓑ Ⓒ Ⓓ Ⓔ
18. Ⓐ Ⓑ Ⓒ Ⓓ Ⓔ   38. Ⓐ Ⓑ Ⓒ Ⓓ Ⓔ
19. Ⓐ Ⓑ Ⓒ Ⓓ Ⓔ   39. Ⓐ Ⓑ Ⓒ Ⓓ Ⓔ
20. Ⓐ Ⓑ Ⓒ Ⓓ Ⓔ   40. Ⓐ Ⓑ Ⓒ Ⓓ Ⓔ

HANDBOOK of MATHEMATICAL, SCIENTIFIC, & ENGINEERING FORMULAS, FUNCTIONS, GRAPHS, TABLES, & TRANSFORMS

RESEARCH & EDUCATION ASSOCIATION

A particularly useful reference for those in math, science, engineering and other technical fields. Includes the most-often used formulas, tables, transforms, functions, and graphs which are needed as tools in solving problems. The entire field of special functions is also covered. A large amount of scientific data which is often of interest to scientists and engineers has been included.

Available at your local bookstore or order directly from us by sending in coupon below.